THE MIRACULOUS PROPERTIES OF IONIZED WATER

THE DEFINITIVE GUIDE TO THE WORLD'S HEALTHIEST SUBSTANCE

By

Bob McCauley, CNC, MH

This book is dedicated to my family, Minsok Kim, my loyal staff and all those who believed in Ionized Water when no one had ever heard of it. To my boys who are always there for me. And to Don Strayer who first introduced me to Ionized Water in 1996. ~ RFM

Health books require medical disclaimers because if you don't provide them you become liable in every way. I don't have any problem stating that I know almost nothing about medicine and have no formal training in the health field. But then, what does it take to understand how to Achieve Great Health, a doctorate or a few letters behind your name? I have learned what I know about health by reading and talking to others about health. More than anything it takes understanding through experience and that is what I have done. These are simply my observations about why I believe *Ionized Water* is so healthy. Anyway, medical and legal disclaimers are necessary so here is mine:

The purpose of this book is to educate. It is sold with the understanding that the publisher and the author shall have neither liability nor responsibility for any injury caused or alleged to be caused by the information contained in this book. This book is not intended in any way to serve as a replacement for professional medical advice. Rather, it is meant to demonstrate that aging can be slowed and even reversed and that Great Health is achieved when the most fundamental nutritional needs of the human body are met. If you feel the need, always consult a doctor or another medical professional when you have an illness or disease of any kind. We admit to knowing little if anything about medicine and therefore would never offer medical advice to anyone for any reason. Medicine is in one direction, health is in the other. The author offers health advice that is his personal opinion only.

The United States FDA has not reviewed ionized water and therefore has made no determinations or assessments about it.

It is hereby declared that the 9th and 10th Amendments to the US Constitution and those rights so granted to Americans by them fully apply to this book.[1]

Note: *Ionized Water* and other key terms are italicized in this book for emphasis.

Edited by: Patti McDowell; Catherine Mack; Tina Rappaport
Cover Design: Bob McCauley, CNC, MH

ISBN: 0-9703933-2-6
Third Edition – May 24, 2010 - RR

Published and Distributed by SE, Inc.
A family owned corporation.

Table of Contents

Preface

"All great ideas are controversial or have been at one time."

~ George Seldes

This book was created using articles I've written on *Ionized Water* as well as sections of my first book Confessions of a Body Builder, Rejuvenating the Body in Spirulina, Chlorella, Raw Foods and Ionized Water[2] and my second book, Achieving Great Health[3]. These works promote *Ionized Water* as the healthiest substance we can possibly put in our bodies. Since 1996 I have done more to promote *Ionized Water* than any other person in North America. I can make that claim because both my books have focused on *Ionized Water* and describe it in simple, everyday laymen's terms that the average person can understand. I knew it was important to write to this level because like most people, I don't have much of a formal scientific background. However, I do have a great love of science and a desire to understand the properties of *Ionized Water*. The ionization of water and the minerals it contains can easily be confirmed through scientific methods and measured using scientific instrumentation.

A group of people who did not entirely understand what *Ionized Water* was introduced me to it several years ago. They knew it was alkaline and that we must put alkaline substances into our bodies if we expect to prevent and cure disease.[4] But they didn't know what made *Ionized Water* alkaline. They knew *Ionized Water* was very hydrating and detoxifying and that it was called micro-water because it contained smaller water molecule clusters, but they didn't know exactly how these tiny clusters were created or why they were important. They knew that *Ionized Water* was not only an antioxidant, but an extremely powerful and effective one. They told me that it was a *free radical scavenger* (Appendix 6), but they did not realize what qualities it possessed that qualified it as one. Nor did they know that *Ionized Water* had a negative charge or *Oxidation Reduction Potential* (ORP)[5], which is a measurement of the millivoltage (mV) of the water. ORP is the most important term to learn regarding health and they didn't even mention it because they hadn't been told that ionization causes a radical change in the ORP of the water. Most of all, they did not understand how *Ionized Water* was created because if they did it would have answered most of the other questions they had as to why it possessed these qualities.

Ionized Water had not been sufficiently explained to me in terms that I, the average person, could understand. I decided to fly to California and visit the importers of Japanese water ionizer models and I learned that the Japanese manufacturers were trying to market their product from Japan using their marketing techniques. I knew this would never work in the American marketplace since Westerners do not think like Asians at all when it comes to health. Westerners historically follow allopathic medicine, which tries to overcome the natural mechanisms of the human body. Asians historically follow naturopathic or holistic medicine, which encourages us to flow with nature, not oppose or try to outsmart it. They also stress preventative health rather than reacting to a disease that has already developed. It is 100 times more difficult to cure the body of any disease than it is to prevent that disease in the first place. Western allopathic medicine attempts to create synthetic substances known as pharmaceuticals to use as weapons to overcome disease when in fact they only control the symptoms of disease. And they do not even do a good job at that because every drug has side effects, some of them quite serious. The medical profession was the third leading cause of death in 2002.[6] The first and second causes of death were cardiovascular disease and cancer. 106,000 people died from the negative effects of pharmaceuticals;[7] 7,000 died due to medication errors in hospitals.[8]

We need to approach the concept of *Ionized Water* with the understanding that it is a phenomenon created in nature, however, here we are able to intensify and concentrate its incredible properties. *Ionized Water* helps the body to flow with nature, not oppose it. This is a lesson that we are still learning in the West. *Ionized Water* was accepted by Asians first because it is found in nature and therefore it is a natural substance that enhances our health. At its core, Asian philosophy has tenets that readily accept the concept of *Ionized Water* because it is completely natural.

I began to examine various water ionizer manufacturers and found there were only a few in Asia. Rumors surrounded *Ionized Water* and what its properties could actually achieve. There were stories that over 25% of all Japanese drank *Ionized Water*, when in fact only 4 – 8% of the Japanese population consumes it regularly. Other stories suggested that it was used extensively throughout Japanese industry, as well as in the agricultural industry for livestock and as a pesticide on golf courses. I have found substantiation for some of these claims, but it is not as widely used as was rumored. *Ionized Water* could potentially be used in numerous capacities in many industries.

Virtually all the water ionizers sold in North America at that time were manufactured in Japan. The first water ionizers in North America were priced at more than US$5000 in the 1980's. *Ionized Water* was completely unknown, in part because it was such an alien concept to the general public. I began importing water ionizers from Korea in 1996 and quickly learned that everything depended on the quality and reliability of the manufacturer's product. My first importation of water ionizers resulted in a 80% failure rate, mostly because of cheap parts. It took a while, but once I had a reliable product, all I needed to do was learn how to market them. My main goal was to bring the price down to make them affordable to as many people as possible. I knew it would be word of mouth that made *Ionized Water* a household name. It was going to be a grass-roots phenomenon, people telling neighbors, friends and family about their wonderful experience with *Ionized Water* that was going to mainstream it into the general population.

I have been blamed for cheapening the value of *Ionized Water* by reducing the price of water ionizers when in fact it has had just the opposite effect. Bringing the price of water ionizers down has resulted in more people having access to *Ionized Water* and becoming healthier by drinking it. They have also spread the word about *Ionized Water* because when someone tries something that makes them feel great, they tell everyone they know about it. For that, I am quite proud of it. Giving people the information they need to become healthier has become for me the most satisfying, rewarding occupation I could ever imagined having.

I have sold thousands of water ionizers over the last fifteen years and I know that it is only the beginning for what will one day become a household word: *Ionized Water*. The water ionizer will one day be as common in the home as the toaster or microwave. It will take much more education and that's what this book is all about: educating the general population that knows nothing about the world's healthiest substance.

I began selling more water ionizers than any other company in the United States within six months simply on the basis of price and reliability of the water ionizers I sold. Since no one knew anything about *Ionized Water*, I began to produce flyers and information sheets that introduced people to the concept of ionization and what they could expect from drinking *Ionized Water*. At the time, the only other book available that even mentioned *Ionized Water* was Reverse Aging by Sang Whang. But there still was almost nothing that described the Miraculous Properties of *Ionized Water* in simple layman's terms that the average person, one not particularly interested in bettering their health, could read and easily understand. Hopefully, they would see that adding *Ionized Water* to their lifestyle could dramatically improve their health.

Around the same time that I found *Ionized Water*, I also discovered *chlorella* and *spirulina*, the most powerful, nutritionally dense whole foods in the world. I developed flyers and information sheets about those as well and began including them with all the water ionizers and other products we were selling. Collectively, those turned into a catalog and eventually I wrote Confessions of a Body Builder, which is not about weightlifting, but building the body with water and nutrients. It was the first book in print that explained *Ionized Water* in simple layman's terms that the average person could comprehend. At present, only a small percentage of the public knows anything about *Ionized Water*, but it is growing in awareness, building momentum as a parallel to the natural health movement. People are beginning to realize that medicine addresses the symptoms of disease; it does not heal the body.

The body can cure all disease if it is sufficiently alkalized, hydrated and detoxified. This will be a recurring theme in this book because it is the recurring theme of health. Alkalization, hydration and detoxification are the basic structural tenets of my health protocol because they are what nature dictates we do if we want to be healthy.

I deliberately use the technique of redundancy in this book because the concepts of ionization, alkalization and detoxification are still quite mystifying to the average person. The use of repetition and describing the same thing in different ways is helpful to someone who knows nothing about what you are trying to teach them. This is especially true when the subject you are teaching is complicated and your reader desperately wants to understand it. It's also why I try to write in the simplest of terms that I possibly can.

I believe the time is right for the West to finally accept and embrace *Ionized Water*. Although it is quite complicated in regard to its mechanics and chemistry, *Ionized Water* is created by a simple process that is easy to understand. I will not discuss its chemistry in detail because I am not a chemist and we do not need to understand the exact chemistry of *Ionized Water* to benefit from it. The same is true for health in general. We do not need to understand the mechanics of human health to be healthy any more than we need to understand how to repair an engine in order to drive a car. All you need to know in order to start a car is how to turn the key. *Ionized Water* is the key we should use to start on the path to Great Health, Perfect Health – the kind of Health *most* people only dream of.

Chapter One

The Water that Becomes Everything

"If it were only the other way! If it were I who was to be always young, and the picture that was to grow old! For that I would give everything! Yes, there is nothing in the whole world I would not give! I would give my soul for that!"

~ Oscar Wilde
The Picture of Dorian Gray

In order to be healthy we need only do three things: *Alkalize, hydrate and detoxify the body*. If we accomplish this, we can prevent and even cure the body of any disease. All we need to know is how to achieve this. *Ionized Water* accomplishes all these things and much more. It *alkalizes, hydrates* and *detoxifies* the body more effectively than any other substance. No other water can accomplish this. Running normal tap water through a water ionizer creates a miracle that can help put your body into a position of health you never imagined you could achieve.

Living Water

"In the world there is nothing more submissive and weak than water. Yet for attacking that which is hard and strong nothing can surpass it."

~ Lao-Tzu

Ionized Water is not only the best water we can drink, it is the best substance we can possibly put in our body. Consumption of *Ionized Water* is critical if we wish to bring the body into balance, a state known as *homeostasis*. Fresh and strong is the best way to drink *Ionized Water* once you've become acclimated to it. How long that will take depends on your overall health and toxicity.

★ *Ionized Water* is negatively charged and alive with electrons, for which our bodies are starved. Along with its alkalizing and hydrating properties, *Ionized Water* is a liquid antioxidant, which is why it can be considered the best substance we can put in the body. I have become younger by drinking it and you will too.

A Short History of Ionized Water

"It is amazing what can be accomplished when nobody cares about who gets the credit."

~ Robert Yates

Negative ions were first studied in Russia in the 1850's.[9] They are one of the most basic, universal building blocks of human health. Without the presence of negative ions, the body cannot survive. Biological aging is a process in which we feed the body a constant stream of positive ions that slowly erodes our health until we succumb to disease and our organs cease to function. The catch-all-phrase used to describe the quiet death of most elderly people is that they "died of old age". What it actually means is that so many of their organs stopped functioning that it's impossible to determine what actually killed them.

The struggle for health is the effort to get enough negative ions into the body. This is accomplished by eating raw foods and drinking *Ionized Water*. It is impossible for us to be truly healthy while living in a positive ion environment, both inside and outside the body. Given that, it cannot be understated that the discovery and subsequent research into negative ions is one of the great moments in the quest for human health.

The Russians did much of the preliminary research relating to the concept and development of *Ionized Water* starting in the early part of the 20th century. They deduced through their discovery and subsequent research of negative ions that since they were good for human health, then positive ions must be bad for human health, which they are indeed. They also postulated that if they could create negative ions in water, then theoretically it could take on certain properties that would be quite healthy. They experimented with simple electrolysis where an anode and cathode are placed into water at the same time creating positive and negative ions that cancel each other out. The challenge they faced was how to separate the ions. The first water ionizers were invented in Russia in the early 1900's. *Ionized Water* is one of Russia's greatest scientific achievements and a tremendous gift to the world for which we should all be indebted, although it has gone almost entirely unrecognized until now.

Further experiments were run on *Ionized Water* in Russia (then the Soviet Union) in the 1940's. In the late 1960's, the Chelyabinsk Project was established to determine if *Ionized Water* could remove radiation from the body, which it effectively does. The Chelyabinsk Project, headed by Vladimir Egov, was named after the small town in the Ural Mountains where it took place. Many people in the former Soviet Union had been exposed to radiation during nuclear research and development at the time of the Cold War. Thus there was a pressing need for methods of removing radiation from the body. However, three effective methods for radiation removal overlooked by the Russians are *Chlorella Pyrenoidosa,*[10] *Focoidan,*[11] and use of *Far Infrared Waves* (FIR).[12] These all remove heavy metals from the body as well, which is facilitated with the consistent use of *Ionized Water* since hydrated tissue is more easily detoxified than dehydrated tissue.

Early water ionizers were developed in Japan during the 1950's. Some of the first experiments were done on plants and animals during this time. The Japanese medical community quickly determined that *Ionized Water* was not detrimental to human health. The first residential water ionizer was developed and sold in the 1958. On January 15, 1966, the Japanese Health Ministry approved the water ionizer as a "Health Improving Medical Device." Korea followed suit and approved the water ionizer as a medical device 10 years later.

During the 1970's and 80's, the research and development of *Ionized Water* continued in other countries, primarily Japan where the first residential water ionizer was introduced. It was the beginning of the domestication of the water ionizer. Becoming healthier would never be easier and the notion of what healthy water is would never be the same.

Debunkers of Ionized Water

"Beware of false knowledge. It is more dangerous than ignorance."

~ George Bernard Shaw

I am considered a "crank" by some, which is aptly defined by Mark Twain as "*a person with a new idea until it succeeds.*" Without fail, there are those who attempt to debunk great advances in every scientific field and *Ionized Water* is no exception. Debunkers suggest that the ionization of water simply isn't possible. One debunker states that *Ionized Water* should be dismissed simply because *it is too good to be true*. Others state that ionization is impossible to verify when in fact its properties are demonstrated using scientific measuring devices such as pH and ORP meters because the changes that ionization produce are radical, immediate and measurable.

Debunkers contend that most new health technologies developed outside of the traditional medical establishment are worthless and have no value beyond their hype. They claim that nearly every unusual invention that doesn't easily fit into the framework of conventional thinking turns out to be yet another urban myth that needs to be debunked. And if something that they have denounced turns out to have merit, not another word is ever said about it and the debunker's web page is quietly removed.

What health debunkers do best is plant doubt in our minds when we are trying to search out the truth about breakthrough health devices. It's easy to be a skeptic. It's difficult to blaze trails and fly in the face of every conventional thought and precept science has established. It's hard to believe in something that you discover and want to be true when someone with a scientific background who claims to be an expert is soundly denouncing it. It is even harder to believe in things that others are ridiculing. Debunkers have a chilling effect on open debate because of the way they often scoff at the technologies they are challenging. Their attack on new health technologies that have not been conducted, sanitized and approved by the medical establishment often extends to the promoter of the new invention. They become the focus of derision much the way Galileo was when he claimed the Sun was the center of the Solar System, not the Earth. Debunkers ridicule those who have found a new path, uncovered a new truth and consequentially have surpassed an old belief system. The water ionizer is a prime example of this kind of attack that stifles progress.

We tend to accept the denunciations of the debunkers because we want to believe that we can't be so easily fooled. We want to deem ourselves too savvy to be deceived by promises that sound too good to be true. We enjoy hearing that ideas and inventions that may sound a little weird, or too good to be true, could not possibly be true. Debunkers want us to believe that life is too harsh for good things to be true. Life is too difficult for something like *Ionized Water* to be healthy for us and possess the miraculous health properties it is said to have. And since someone with a scientific background has claimed it to be a hoax, it must be one. Thus, we doubt and dismiss it, exactly what debunkers want us to do.

Debunkers tend to be heavy on opinion and light on the facts. And they often present their opinion as fact. They are sometimes chemists or chemistry professors who understand book chemistry, but don't seem to comprehend the simple chemistry of *Ionized Water*, that of charging water by creating negative ions (-OH) and positive ions (+H). Nor do they understand how these ions are created by the simple process of electrolysis then separated through a membrane. It is not difficult to understand *Ionized Water*, but it is difficult for some to accept that something so simple can have such a profoundly beneficial effect on everything it comes in contact with, especially our health.

I don't expect that debunkers will ever disappear. Ironically, debunkers' websites are a good resource for groundbreaking technologies since they tend to dismiss anything that might possibly work.

The notion of having people around to challenge the traditional conventions and new inventions is a good one because there is plenty of fraud in the world. Debunkers, on the other hand, are often something different than those who expose charlatans. Debunkers stand in front of progress by denouncing and ridiculing any radical new concept. However, armchair investigations and conclusions drawn from them will never replace field testing and something debunkers almost never do is conduct their own legitimate scientific investigations.

Debunkers often ply their trade in order to protect the *status quo* and the safety it provides to those who are afraid of the future and the change that it inevitably brings. Debunkers sometimes serve no other purpose than to confuse us regarding complex issues, which is especially disappointing and destructive when it comes to natural health.

When a radical scientific concept is first introduced, there is commonly resistance to it by the prevailing scientific community. New ideas are generally ridiculed and derided because they make people uncomfortable. Scientists don't want to accept new ones because their life's work is based on the foundation of these old ideas. They become particularly defensive and resistant when the new ideas destroy, override or completely supersede the old ones. Often, the members of the established scientific community must die off so the next generation can be raised with the new concept and accept it from the beginning without any unnecessary bias and hand-wringing about what they once embraced as the truth that has now been superseded by the next wave of scientific discovery. With the coming of the next generation, the new concept will no longer be regarded as radical, but rather another stepping stone to understanding the whole truth about health, nature and the very essence of human existence. The acceptance of *Ionized Water* is yet another stepping stone that cannot be ignored because it is a critical component of health.

A considerable amount of research has been conducted on the effects of *Ionized Water*, most of it by Japanese researchers, which should be no surprise since they have embraced the concept of water ionization more than anyone else (Appendix 5: Scientific Studies). All of the research done on *Ionized Water* has arrived at the same conclusion: ***Ionized Water* benefits everything it comes in contact with as long as it's used correctly.**[13] Physical scientific measurements, not conjecture, demonstrate the properties of *Ionized Water*. With this hard evidence in front of us, there is no other conclusion to reach than to claim that the properties of *Ionized Water* are best described as simply miraculous.

The strong **detoxification** effects of *Ionized Water* alone are enough to demonstrate that it is not like any other water you will ever consume. This is due to the small water-molecule cluster size. How water molecules group together is another thing that debunkers claim can not happen, thus it is so much more quackery. This conclusion is reached despite the fact that a substantial amount of scientific research has been done on water molecule clusters.[14] Water-molecule cluster size is measured by using a Nuclear Magnetic Resonance device. We can also determine the change in the surface tension of *Ionized Water*, which is measured in dynes.[15] A lower surface tension demonstrates the existence of smaller water molecule clusters.[16]

Debunkers do not accept the basic concept of positive and negative ions and how they affect human health. If they did, they would understand how important it is that we put alkaline substances in our body. Those who would attack *Ionized Water* and claim it is a hoax do not understand the value of consuming substances that are negatively charged and possess negative ions. By their own admission, those who debunk *Ionized Water* have never consumed it, claiming there is no need to because the concept of ionizing water is impossible to begin with.

Debunkers are part of a faux-intelligentsia that does not believe in field testing before they arrive at conclusions. The truth is that field testing often reveals many things that formulas and theories written on paper could never have predicted. Examine debunkers and their methods closely for they are self-appointed watchdogs. The methods of debunkers are never based on scientific evidence, but rather their own personal interpretation of the science they are debunking. Debunkers, more often than not, need themselves to be debunked. This has never been truer than when it comes to the debunkers of *Ionized Water*. They don't understand the simple truth that ionizing water takes it to a level of health many times above that of conventional water.

Chapter 2

Water - Where Life Begins

"I believe that water is the only drink for a wise man."
~ Henry David Thoreau

The person who is hydrated . . .
is always healthier . . .
is less toxic . . .
is more energetic . . .
functions more efficiently overall. . .
retains less water in the extremities (legs, feet, hands) . . .
is more mentally alert. . .
is less prone to disease . . .
is better able to fight disease of every kind. . .
. . . than the person who isn't.

In the Beginning Was Water

"Begin with the beginning."
~ Lord Byron Don Juan

Health begins with water, for it is the most essential element needed by the body. Water is the cornerstone of health. It is a universal solvent and the body's lubricant. Proper hydration of the body is crucial to human health in countless ways. You will never be truly healthy if you are not sufficiently hydrated. When we begin making water a big part of our lifestyle, we take our first steps toward true health.

All life springs from water. We evolved from the sea. *"When the animals went ashore to take up life on land, they carried part of the sea in their bodies, a heritage which they passed on to their children and which even today links each land animal with its origins in the ancient sea."* [17] Every fish, reptile, animal and mammal carries the elements of sodium, calcium and potassium within their veins in almost the exact same proportions as sea water. We are only a few steps removed from today's creatures of the sea.

Bones and the Marrow of Life

"To get back my youth I would do anything in the world, except take exercise, get up early, or be respectable."

~ Oscar Wilde
The Picture of Dorian Gray

We are born as water babies. We spend the first nine months of our lives in the embryonic fluid of our mother's womb. It is as though we have emerged from the sea at the moment of birth to take our first breath of oxygen. A baby's bones are mostly water at birth. They are much more pliable than any other time in our lives. As we age, they begin to dry out until they become brittle. They break more easily as they become parched.

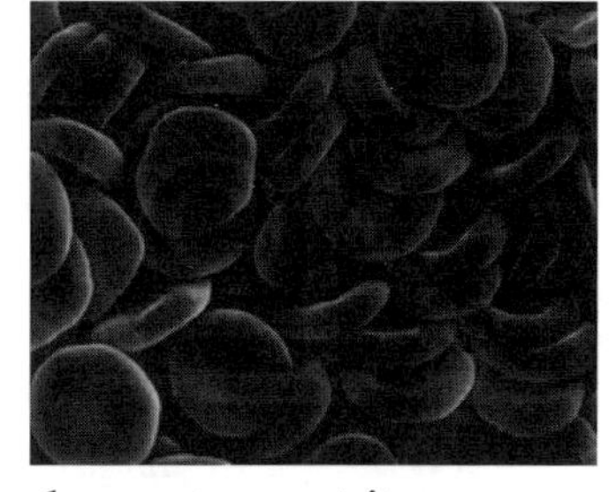

The body cannibalizes the calcium from the bones to meet its requirements for the rest of the body. By retirement, the average senior citizen has become 50 – 70% dehydrated, which is why nearly all elderly people have every imaginable disease, including nearly ubiquitous constipation throughout the age group.

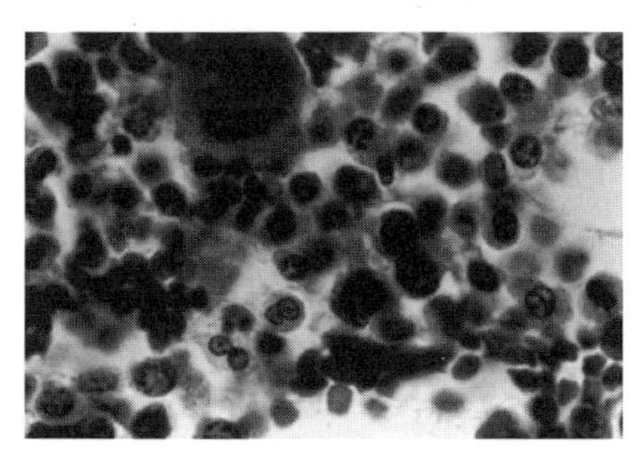

Bone marrow is a major component of our immune system because it produces our T-cells, B-cells and other cells that collectively form our immune system. These white blood cells produce antibodies and have a myriad of other immune functions. Red blood cells carry oxygen and produce stem cells and blood platelets that allow the blood to clot. Stem cells are primitive, unprogramed cells that can continue dividing forever and can be formed into any other cell needed by the body. Our bone marrow is the womb of these critical human cells, and as it becomes increasingly desiccated by chronic dehydration it loses its ability to produce them, which encourages disease, structural decline of the body and accelerates the aging process. The importance of bone marrow to human health cannot be underestimated. Keeping our bone marrow constantly hydrated, cleansed and rejuvenated is one of the most critical components of Great Health and longevity.

Water: The Most Important Element for the Body

"Trickling water, if not stopped, will become a mighty river."

~ Chinese Proverb

True health cannot occur without proper hydration of the body. Every organ in the body heavily depends on water to function properly and to its capacity. We are mostly water. The average human body is 69% water. The brain is 85% water, bones 35% water, blood 83% water and the liver 90% water. When we become dehydrated, we put our health in immediate jeopardy. Thus, we gamble with our lives without realizing the dangerous high-wire we are walking when we don't drink enough water.

We must drink half our body weight in ounces minimum each day. For instance, if you weigh 200 lbs, you should consume 100 ounces of water each day. However, I recommend people drink a lot more because we lose that amount of water through the basic functions of the human body, those of urination, perspiration, respiration and defecation.

The National Research Council guidelines suggest we require 1 milliliter of water for every calorie of food consumed, which is approximately half our body weight in ounces. However, their official report states that *"most of this quantity is contained in prepared foods"*, implying that you don't need to drink water because you will get it from the foods you eat. These types of misconceptions cause many of us to remain complacent and conditioned with the notion that water is more a choice than necessity. Digestion is a dehydrating process that requires a huge amount of water from the body's reserves. Also, the typical diet consists of dry, cooked and processed foods, unlike *raw foods* that contain significant amounts of water.

Some have concluded that the water found in *raw foods* is sufficient to properly hydrate the body. This is not true, but even if it were, we still need to consume large amounts of water to flush our digestive tract between meals, which is one of the healthiest things we can do. Adopting the practice is much more of a necessity than choice than most of us realize. None of the myriad processes of the body will function to their capacity if the body is not sufficiently hydrated. Water is required in every action, reaction and bodily function even though many prominent naturalists[18] and raw foodists barely make mention of it.

Revealing the True Healer: Water vs. Medicine

"A great pleasure in life is
doing what people say you cannot do."

~ Walter Bagehot

Traditionally, water has been considered mere packing material that serves little purpose other than to give the body its weight and bulk. Medieval thinkers still believe water is there to give the body volume, otherwise it would be nothing except dry chemicals. To this day, the medical establishment sees water as little else when in fact water serves to energize every cell and organ in the body. It is crucial to every bodily operation and when we become dehydrated, the body instinctively begins to ration water to each organ. The brain, being the most important organ, gets the most water. The skin, being the least important, is rationed the least amount of water. Chronically dry skin and/or dandruff are signs of advanced dehydration, as are asthma and hyperventilation. It is the law of *vital adaptation* at work. The body will do what it must to survive. It will adapt to its circumstances, which in this case means the most important organs such as the brain, heart and liver get water first. If the body didn't do that we would suffer the ill effects of dehydration much more rapidly.

Exercising provides far greater benefit when the body is properly hydrated. Researchers have determined that 1% dehydration causes a 5% drop in athletic performance.[19] If we desire peak performance, we need to hydrate ourselves and that can only happen by consistently drinking water throughout each day. Water energizes the body and brings it to life. I always hydrate myself with *Ionized Water* just before exercising.

Liquid of Life

"The art of medicine consists in amusing the patient while nature cures the disease."

~ Voltaire

It is quite difficult for the human body to get water from any other source other than water itself, yet so many of us, including certified nutritionists and dietitians, believe that simply because a substance contains liquid it will provide the body with water. Nothing could be further from the truth. Some of the most dehydrating substances for the human body are mostly liquid. Most people reach for a caffeinated beverage either coffee or soft drinks to wake and energize themselves. Rather than make us mentally alert, caffeine temporarily stimulates the body's nervous system. When it wears off, we deflate and crash. Stimulants of any kind are a great detriment to the body. The first thing to reach for when we wake in the morning is a tall glass of water, preferably *Ionized Water*.

What is important to know about the liquids we consume is what other substances they contain besides water because certain substances can cause tremendous dehydration of the body.

Tasty, Refreshing Poison

"Poisoners they are, whether they know it or not."

~ Friedrich Nietzsche

Carbonated soft drinks are primarily water, but they are the most dehydrating substances that are commonly consumed. I grew up on a steady diet of soft drinks and it did more damage to my health than anything else. As a child, I was never warned how dangerous soft drinks were to my health, and at the time few questioned them. Not surprisingly, many CEO's of soft drink and fast food chains succumb to cancer and heart disease at relatively young ages.[20]

"Consuming acidic foods such as soft drinks may also create an ideal environment for cancer to form."[21]

Soft drinks contain highly processed sweeteners, which do not remotely resemble the natural chemical structure of sugars found in fruits and vegetables. Caffeine is a diuretic that has been linked to rheumatoid arthritis.[22] It encourages the kidneys to produce so much urine that it leads to dehydration. The phosphoric and carbonic acids contained in soft drinks are extremely acidic (pH 2.7)[23] and cause the body's pH to drop considerably, leaving us more susceptible to chronic, bacterial and viral disease of every kind. The phosphates found in soft drinks block the absorption of calcium and leach alkaline minerals from the body, leading to bone loss.[24] Soft drinks also severely erode our teeth.[25] They contain large amounts of benzene, eight times that which is allowed in drinking water.[26] Benzene is a carcinogenic chemical that has been linked to leukemia. Chronic soft drink consumption was linked to hyperactivity and mental health problems amongst teens in a Norwegian study of more than 5000 teens.[27]

"To prevent acidic poisoning from cola (or other acid) consumption, the body use two strategies. One is to use alkaline blood buffers to buffer the acid. The other strategy is to convert these volatile liquid acids into less-reactive solid acids."[28]

The dangerous nature of soft drinks comes with no warning about their potential damage to our health and one of the reasons is that medical, educational, health and fitness professionals regularly promote the consumption of soft drinks. If the medical establishment was truly concerned about our health, they wouldn't allow the sale of soft drinks in hospitals, for example. Instead, we do not hear a single word of condemnation about soft drinks from them.

Dr. Heinz Valtin[29] stated: "*Many people contend that caffeinated drinks such as tea and soda, don't count toward your fluid intake. In fact, they can. And thirst does not necessarily mean you're already dehydrated."*[30] Both of these statements could not be further from the truth, however the simple fact that this appeared in a newspaper makes it the "truth" to many people. A medical doctor has approved it, so now their concerns over soft drinks being unhealthy for them are eased. They have found an enabler who has rubber-stamped their rationalization of why they consume soft drinks and now they can swing it like a mighty club of defiance. The fact is that by the time the thirst mechanism in the body switches on to tell you to drink water, the body is already dehydrated.

Most of us instinctively know that soft drinks are bad for our health even though the yearly per capita consumption of soft drinks in the U.S. has risen steadily to 49 gallons over the past 50 years.[31] This is in large part due to marketing and the lack of condemnation from parents, school officials and other role models where inculcation to the soft drink culture often starts as early as one year old.[32]

Soft drink consumption has been linked to esophageal cancer.[33] The artificial sweetener aspartame found in diet soft drinks has been linked to birth defects, brain cancer,[34] emotional disorders, carpel tunnel,[35] diabetes and epilepsy. When asked if they would allow a product such as soft drinks on the market, the Food & Drug Administration (FDA) was blunt in stating that there was no possibility they would allow it to be marketed if it were presented to them today for approval. Soft drinks are poison and should be avoided at all costs (See Appendix 2).

The Crippling Results of Dehydration

"Life is not complex. We are complex. Life is simple, and the simple thing is the right thing."

~ Oscar Wilde

Psychiatric illnesses such as Alzheimer's Disease are a sure sign of a lifetime of chronic dehydration. Every time you take a sip of water, it is dedicated to hydrating the brain and surrounding cranial fluids first, then to the rest of the body. When it is sufficiently hydrated, it functions closer to its full capacity. When we fail to drink enough water over a lifetime, toxins begin to interfere with the brain's delicate neurochemical balance. Once a single chemical reaction in the brain is significantly altered, the resulting ripple effect will be felt throughout the brain. The accumulation of these events can lead to brain diseases and imbalances of all kinds, including clinical depression, schizophrenia, bipolar disorder, Lewy Body disease and other dementia-related diseases. Reversing these psychiatric conditions is not easy, but can be done by following a strict 100% diet of raw foods, using far infrared rays and drinking 1 – 2 gallons of *Ionized Water* each day. However, people who suffer from dementia are often so disconnected from reality that getting them to actually change their diet and drink that much water would almost be an impossible task.

Constriction of blood vessels and capillaries can also lead to various dementia-related diseases. Blood vessels collapse and constriction is another result of chronic dehydration.

Doctors often misdiagnose patients because dehydration of the organs results in symptoms that mislead the average medical professional who isn't trained to recognize dehydration or the importance of water to the human body. In fact, water is barely mentioned in medical school. If people were to simply drink enough water and keep their bodies properly hydrated, 60 – 80% of the chronic disease in our society would be seriously abated if not completely absent. When it comes to health there is no substitute for drinking sufficient amounts of water each day.

Too Much of a Good Thing?

"Sometimes too much to drink is barely enough."

~ Mark Twain

The most important piece of health advice I will ever offer anyone is to drink plenty of water! The first line of defense against disease is a properly hydrated body. Water is the best substance we can put in our bodies and should be consumed at all times of the day except around meal time.

The most spring, mineral or tap water we should consume in one hour is one liter, or 1.5 liters of *Ionized Water*. Drinking one gallon or more of water in one hour is far more than the body can process and we essentially drown ourselves. Since the body can't rid water that quickly, the excess goes to the bowels, which pulls salt into it, diluting the concentration of sodium in the tissues, including the brain.[36] This condition is referred to as *hyponatremia* or very low blood sodium. Changing the concentration of salt in this way causes a sudden shift in bodily fluids. This can induce swelling of the brain, leading to possible brain damage.

Over-hydration can also occur if the body is depleted of its vital electrolytes: sodium, calcium, potassium and magnesium. Complete depletion of electrolytes of the body would result in death by a series of cascading events that lead to organs shutting down. Both enzymes and electrolytes are not only essential, they are critical to our health on a moment-to-moment basis. Electrolyte depletion can easily occur during long athletic events such as a marathon if people run 2 – 5 hours without replacing the tremendous loss of water and electrolyte salts they experience. Marathoners put their bodies to the test and drain it of its vital resources like few endurance events can. If you want to see yourself 20 years in the future, look at your face in the mirror the morning after a marathon. Many amateurs who run in marathons do not properly condition themselves to put their bodies through such tremendous stress. To do so and survive, let alone remain healthy, takes a professional whose body is sufficiently conditioned to deal with the incredible stress of dehydration and electrolyte depletion. Those who run marathons and do not regularly replenish their electrolytes put themselves at great risk. Several people have become seriously ill running marathons and a few people have even died. *Hyponatremia* is a rare condition.

Running a marathon is a tremendously strenuous act on the body. Running full-steam for that period of time can put the best of us in danger if we are not careful. Yet some will use it as a reason why we should limit our consumption of water to only two liters per day, which is absurd and counterproductive to our health.

Under normal everyday conditions, the average adult requires 1.5 – 2.0 gallons of water each day, consumed between meals, if they expect to be truly healthy. One gallon of water each day will only meet our basic hydration and internal cleansing needs, while consuming half our weight in ounces only replenishes the amount of water we lose daily. The extra is used to flush our digestive tract between meals.

Sport drinks contain high amounts of sugar. The electrolytes they contain are not easily absorbed by the body because they are not ionic.[37] Thus, they leave us with the false impression that they will adequately provide the body with electrolytes when in fact they do not. I get my electrolytes from angstrom minerals (atomic size), raw foods, unprocessed Celtic sea salt and sea vegetables (Nori, Dulse, Kelp, etc.). The most important whole foods I consume that have electrolytes are spirulina and chlorella. Spirulina is produced with both fresh water and sea water, so it contains all the minerals that are in sea water, including the electrolytes we require. Marathoners who include angstrom minerals, unprocessed sea salt and these raw whole foods in their diet will not have trouble with electrolyte depletion and *hyponatremia* when running such a long distance, dehydrating event.

Chapter 3

What Kind of Water

"The strongest principle of growth lies in the human choice."

~ George Eliot

Which water you drink is of great importance. Thirty years ago, the bottled water industry barely existed in North America. Since then it has burgeoned into a multi-billion dollar business. This has occurred in large part because of the greater awareness of the importance of water to human health. Now that this has been established and accepted by the general public, the debate continues over which water is best to drink. One of the purposes of this book is to once and for all settle that debate.

Bottled Water

Inside my empty bottle I was constructing a lighthouse while all the others were making ships."

~ Charles Simic

I have been in the bottled water business since 1993. What very few people who are in the bottled water business realize is that they are really in the health business because the number one reason people drink bottled water is for their health. Since I understood what business I was in, it was a natural fit for me to start selling *Ionized Water* since it, too, is a health product.

There are hundreds of bottled waters on the market, most of them not any better quality than what comes out of your tap, although they sell for up to hundreds of times the price of tap water. The cost of tap water averages about one cent per gallon. Bottled water sells for .99 cents to $16.00 per gallon depending on the brand and bottle size. Although most bottled water is acceptable to consume, there is nothing in any of these bottled waters that would make it worth these high prices.

2.7 million tons of plastic are used to bottle water each year. 1.5 million barrels of oil are used to make the polyethylene terephthalate (PET) bottle, which is enough to fill the fuel tanks of 100,000 cars. Toxic chemicals are released into the environment during the production and disposal of these bottles. 86% of PET bottles end up in landfills and it takes thousands of years for them to degrade. The practice of tapping ground water supplies for bottled water has led to shortages in several places such as India.[38]

Purified water is acidic. Expensive foreign bottled water brands are not worth the money. Shipping water any distance is quite costly. Shipping water half way around the world is ridiculous from an economic prospective.

If you buy bottled water, make sure it's spring or mineral water that has total dissolved solids (TDS)[39] of at least 100 ppm (mg/l). Water below 50 ppm TDS should not be consumed because it will leach vital minerals from the body.

Much more suspect is surface water, which is drawn from rivers, lakes and reservoirs and is vulnerable to air pollution, agricultural and other industrial runoff. This water definitely needs to be filtered before drinking, although certain elements such as petrochemicals cannot be filtered out and must be removed by purification methods such as reverse osmosis and distillation, which renders the water undrinkable unless it is re-mineralized.

pH of Various Waters

All liquids possess a pH. The pH of the liquids we consume is of great importance. We should consume only alkaline substances such as raw fruits and vegetables and *Ionized Water* if we want to be healthy. Below is a list of various types of water and other beverages and their average pH range.

- Carbonated Water: 4.0 – 5.0
- Imported Carbonated Spring Water: 4.0 – 4.8
- Distilled Water: 5.8 – 6.5
- Reverse Osmosis Water: 5.5 – 6.3
- Bottled Spring or Mineral Water: 6.0 – 8.0
- Municipal Water: 6.7 – 7.1
- *Ionized Water*: 8.0 – 9.9*
- Orange Juice (pasteurized/bottled): 5.0
- Wine: 5.5
- Beer: 4.0
- Liqueur: 4.0
- Whiskey and other distilled spirits: 2.5
- Coffee: 5.0
- Black Tea: 5.5
- Green or Oolong Tea: 6.5

*The pH range of *Ionized Water* is adjustable.

Purified Water and its Dangers

"If you prick us do we not bleed? If you tickle us do we not laugh? If you poison us do we not die? And if you wrong us shall we not revenge?"

~ William Shakespeare

Water cures, but purified water is a detriment to the health of anyone who consumes it. Purified water is produced by *deionization, distillation* or *reverse osmosis* and should not be consumed for three reasons. It acidifies the body, leaches minerals from the body and the large size and shape of its water molecule clusters do not hydrate the body well. In fact, long-term use of purified water can leave us dehydrated.

Purified water should be used in humidifiers, fountains, autoclaves and fish tanks. It is used in many industries, but it should not be consumed.

Purified water has had all the minerals removed from it, which is a substance that cannot be found in nature. When we drink purified water, we are consuming a pure chemical substance, H_2O. Pure chemical substances of any kind are not found in nature collected together, but instead are always mixed with other substances. Because of its extreme purity, purified water absorbs carbon dioxide from the air, which makes it acidic and even more aggressive at dissolving the alkaline substances it comes in contact with. The pH of purified water is much more acid than the body should be. It contains little or no dissolved oxygen and therefore is considered dead water unless it is re-oxygenated.

Purified water enters the body pure, but does not come out pure. It leaches vital minerals from the body, turning it more acidic. Purified water that has been ionized should also be avoided. Even though ionization does increase purified water's hydration properties, the trade off is not worth it because it still steals vital minerals from the body due to its acidity and absolute purity.

Ionization will change the pH of purified water temporarily even though there are no minerals in the water and almost no conductivity. Without conductivity, the change in pH of purified water will be short-lived compared to water that contains minerals because they act as a conductor. The ionization process steals * electrons from one water molecule and donates it to another water molecule causing a change in the water's pH value. Since the ionization of purified water is weak, the water quickly reverts to its original condition since there is nothing to hold, or bond, the extra electrons in place, thus they immediately dissipate.

Dr. Masaru Emoto[40], a Japanese researcher, demonstrated that water which had been distilled, polluted or passed through the body after consumption had lost its structure or *inner order*. Destruction of this natural crystalline structure renders purified water useless to the body. It has been changed beyond its innate composition and structure the same way cooking foods changes their natural chemical structure. Heat destroys the energy of raw foods by destroying its enzymes.

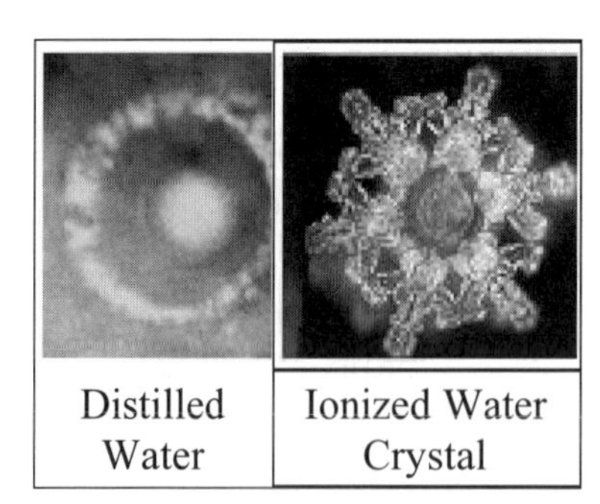

Distilled Water	Ionized Water Crystal

Purified water proponents insist that rain water is identical to purified water because rain is condensed from clouds which are accumulated air moisture. Rain water, however, has a crystalline structure when frozen called snow, demonstrating that it is completely different from purified water which has no crystalline structure when frozen because there are no minerals for the water molecules to form crystals around. The fact that purified water has no crystalline structure when frozen speaks volumes about the characteristics of purified water because it demonstrates that it has no frequency or vibration the way all things naturally have. This micro-vibration is imperceptible to us, but it is everywhere.

The size and shape of purified water molecule clusters is substantially different from conventional water, rain water or *Ionized Water*. Water has been shown to have a "memory," meaning that even if it is substantially changed it will eventually revert to its original shape and vibration frequency. However, if water is completely stripped of its minerals and oxygen, it is dead and this "memory", or former structure, is lost. This same scenario also occurs when water is mixed with toxic pollutants such as chemicals or radiation.

The water found in plants is similar to distilled water in its purity. However, unlike purified water it has smaller water molecule clusters. And when we consume a plant such as a carrot, we do not only consume the water, but the whole plant as well, thus the water is buffered by the alkaline nature of minerals and enzymes contained in the plant. But to extract that water and consume it without neutralizing its purity and acidic nature, our body becomes the buffering agent and alkaline minerals are leached from it.

The list of purified water proponents who are raw foodists and other naturalists is long. Dr. Norman W. Walker,[41] for instance, was a strong proponent of purified water, as was Paul Bragg.[42] I am one of the few naturalists who strongly opposes the consumption of purified water, but I am not entirely alone.

"The longer one drinks purified water, the more likely the development of mineral deficiencies and an acid state. Disease and early death is more likely to be seen with the long term drinking of purified water." [43] Since it has been transformed into an unadulterated chemical substance (H_2O), purified water naturally seeks to balance or mollify its extremely pure, acid condition with an alkaline buffer. Removing minerals from water makes it hungry for anything that is not like itself. All things in nature strive to balance themselves. What is immediately available to purified water when it is consumed are the alkaline minerals and other alkaline material of the body. Thus they are absorbed by the pure water and carried away with it as it leaves the body. It is the extreme characteristics of purified water that encourages this leaching of minerals from the body. Purified water also leaches plastic from the bottles it is contained in, which makes drinking purified water worse when consumed from a plastic container.

"Fasting using distilled water can be dangerous because of the rapid loss of electrolytes (sodium, potassium, chloride) and trace minerals like magnesium, deficiencies of which can cause heart beat irregularities and high blood pressure. Cooking foods in distilled water pulls the minerals out of them and lowers their nutrient value." [44]

One person's black hair turned gray after only two weeks on a distilled-water fast because copper, which is used to form hair pigment, was leached from his body. Years later, his hair still had not fully regained its color and was left 60% more gray than before the distilled water fast. His planned six week fast had to be cut short to two weeks when he began experiencing trembling and short-term memory loss because of electrolyte depletion. While purified water can remove toxins, it will also take with it vital minerals and other elements important to the body, which can promote osteoporosis, osteoarthritis and a host of other diseases since minerals are used in virtually every metabolic process of the body.

Reverse osmosis water is promoted by most health food stores, which is one of the reasons it continues to be touted as the water of choice amongst those who strive to be healthy. Since it has become the "sanctioned water" of most health food stores, convincing people that it is bad for them is made evermore difficult because it flies in the face of an established precept of the monolithic, lockstep natural health industry.

Unfortunately, most water available in convenience stores and vending machines is purified, primarily because the soft drink industry controls so much of the distribution and vending machine territories. Since soft drinks are made with purified water, it is easy for them to bottle and sell what they already produce. To them, water is simply another product in their beverage line, thus no consideration is given to whether it is worthy to consume.

Many will argue that taking mineral supplements can compensate for the theft of minerals due to the consumption of purified water. If you include unprocessed sea salt, spirulina and chlorella, sea vegetables and a large variety of other *raw foods* in your diet, there may be a chance there would be a sufficient amount of minerals and electrolytes in them to replace those that are stolen. But trying to replace them with mineral supplements is nearly impossible, regardless of whether they are angstrom-size or chelated for better absorption. Minerals are most easily absorbed by the body when they come from whole raw foods because they are ionic.

"There is a correlation between purified water and how it steals calcium from the body. [As this happens] the body begins to cannibalize itself to meet the calcium requirements of the body. Chronic digestive problems such as gas and diarrhea are common amongst purified water drinkers. As the consumption of soft water (distilled water is extremely soft) increases, the incidence of cardiovascular disease also increases. Cells, tissues and organs do not like to be dipped in acid and will do anything to buffer this

acidity including the removal of minerals from the skeleton and the manufacture of bicarbonate in the blood." [45]

Naturalist Paavo Airola[46] also warned about the dangers of purified water. Purified water is dead, meaning that it is void of oxygen and has a positive ORP. We should strive to consume substances that have a negative charge, or ORP, such as *Ionized Water* and raw foods.

"The longer one drinks distilled water, the more likely the development of mineral deficiencies and an acid state. I have done well over 3000 mineral evaluations using a combination of blood, urine and hair tests in my practice. Almost without exception, people who consume distilled water exclusively, eventually develop multiple mineral deficiencies." [47]

Proponents of purified water state that the inorganic minerals found in conventional water collect in our veins over time and become like cement, which is an absurd and entirely unsupported notion, scientifically or otherwise. They claim that consuming any water with minerals in it will eventually turn a person into a "stone statue." This claim is a complete myth. Autopsies of elderly people who have consumed water with a high mineral content all their lives have shown no indication whatsoever of cementing in their veins. No one has ever turned into a stone statue, even people who have lived over one hundred years. The deposits that have been found in veins are composed of hardened plaque, cholesterol and nannobacteria, which deposit a calcium shell in order to disguise themselves from bodily defenses that would otherwise destroy them.[48]

"Those who supplement their distilled water intake with trace minerals are not as deficient but still not as adequately nourished in minerals as their non-distilled water drinking counterparts even after several years of mineral supplementation."[49]

Proponents of purified water also claim that the amount of minerals in conventional water is minuscule, about what is found in a slice of orange per eight ounce glass of water. This is true, but misses the point entirely. The amount of minerals found in conventional water is irrelevant. They are in the water to provide better hydration and a crucial buffer that prevents minerals from being leached from the body. Conventional water also does not acidify the body because it contains minerals, although they will not be absorbed by the body because they are inorganic.[50]

Purified Water can be used to detoxify the body on a short-term basis, one or two days at the most. It can leach toxins out of the body quite efficiently, but does so in a completely different way than *Ionized Water,* which is safer and more effective than purified water. I have never done a purified water purge or fast, nor do I recommend them.

"Avoid distilled water as it has the wrong ionization, pH, polarization and oxidation potentials. It will also drain your body of minerals."[51]

The hydration properties of purified water are extremely poor. People who drink as much as one gallon of distilled water per day for years can find themselves dehydrated because purified water does not sufficiently hydrate the body and its cells. One of my customers was informed by his doctor that tests had confirmed he was severely dehydrated even though he'd been consuming a gallon of distilled water each day for over ten years. The structure of a purified water molecule cluster is five-sided or pentagonal, and the angle at which the molecules come together, called the *bond angle,* is extremely large (104 degrees), thus forming a water molecule cluster that is also quite large. The larger the *bond angle* of the water molecule cluster, the less hydrating the water is. These large water molecule clusters do not have the ability to penetrate body tissue or its cells efficiently. This is why purified water does not hydrate the body well and can even leave it dehydrated.

Resizing Water Molecule Clusters

There are machines that spin water the way a blender does. Some of them also magnetize the water. What changes in the water when it is spun is its surface tension, as it does with *Ionized Water.*

This lower surface tension is indicative of smaller water molecule clusters and the presence of more energy, although not enough to be of any significant use to the body. The spinning action does not ionize or significantly change in pH. (For instance, it can only raise the pH of water to 7.0 – 8.0). Running water over quartz crystals will accomplish similar results, making the water slightly more hydrating, but not resulting in ionization or a change in pH value.

Alkalizing Drops

"Books are only the shadow and life the real thing."

~ Esther Forbes

Think of alkalizing drops as books and *Ionized Water* as the real thing. That's the difference between making water more alkaline with certain substances and ionizing water.

Re-mineralizing purified water changes the cluster size, crystalline structure and pH balance, making it safe for consumption. This is accomplished by adding water-soluble minerals such as calcium and magnesium to the source water. There are also catalyzing electrolyte solutions that can be added to purified water that will increase its pH to 9.0 – 10.0 and the TDS (amount of dissolved minerals) of the water to between 50 – 70 ppm, which is the minimum amount of mineral content that water should contain if it is to be consumed. While this dramatically improves purified water, making it fit for human consumption, it is costly and no substitute for *Ionized Water*, which also has antioxidant, superior hydration and detoxification properties.

Adding any alkaline water-soluble mineral such as calcium, magnesium or phosphorous powder to purified water will achieve similar results. Minerals increase the TDS of the water they are added to. It is also less expensive, however, it will not raise the pH of the water the way alkalizing substances catalyzing electrolytes can. They are able to raise the pH so high (pH 10) with a small amount of liquid because purified water is void of minerals and therefore has no absorbent buffer. Since there is no resistance or mineral cushion in purified water, a small amount of these alkalizing liquids are able to dramatically raise the pH of purified water. Similar results can be achieved by adding a small amount of sulfuric acid to purified water, although the pH moves downward to 3.0 or lower.

Bureaucratic Labeling Laws. FDA labeling regulations require that only water purified by reverse osmosis or distillation can be labeled *Drinking Water*. Leave it to regulators and their ignorance to confuse the public on the issue of labeling, but it is crucial. Don't consume bottled water labeled *Purified Water* or *Drinking Water*. It is not fit for human consumption and is dangerous to our health.

Ionized Water vs. Purified Water: When Opposites Meet

Ionized Water and purified water are exactly the opposite from each other in every regard.

- Purified water is acidic; *Ionized Water* is alkaline.
- Purified water does not effectively hydrate the body; *Ionized Water* is extremely hydrating.
- Purified water leaches minerals from the body; *Ionized Water* provides minerals to the body.[52]
- Purified water does not provide the body with oxygen; *Ionized Water* does.
- Purified water does not scavenge for free radicals; *Ionized Water* does scavenge for free radicals.
- Purified water encourages oxidation of the body; *Ionized Water* reduces oxidation.
- Purified water has a positive ORP, which is an oxidant. *Ionized Water* has a negative ORP, which is an antioxidant.

Municipal, Tap or City Water

"When drinking water, think of its source."
~ Chinese Proverb

For a fraction of the cost of bottled water most people can drink the municipal water that comes out of their tap as long as it is properly filtered. Depending on the city, all municipal water comes from natural sources, including ground water, surface reservoirs, lakes and rivers. There are no municipalities in the United States that recycle its sewage water back into drinking water to be pumped out to the community for public consumption. All waste water produced by municipalities discharge to locations where the water will not be reused until nature has had a chance to reclaim and therefore filter it. Municipalities do not return treated waste water to the same reservoir it is drawn from. If it draws water from the same river that it discharges treated waste water to, it does so downstream from its extraction point. The only recycled water consumed in the US is that which has been filtered through the natural recycling processes of the Earth over a period of many years.

The quality of tap water in North America varies greatly, determined primarily by its source, pre-treatment and how large of a community it serves. Small communities that grow suddenly in a short period of time can sometimes have poor water quality because to meet demand the municipalities must continually add chlorine and other chemicals to the water to avoid bacterial contamination. The additional need for chlorine, progressively increased in volume in the water, creates a vicious cycle of requiring more and more chlorine to disinfect the water supply system. Chloramine is also used in municipal water for disinfection.[53]

Since chlorine is a strong oxidant, it oxidizes, deteriorates and ages the body. It is imperative that chlorine be removed from water before it is consumed. Chlorine dries the skin and scalp causing them to flake. It is a known carcinogen[54] and has been linked to the increased risk of stillbirth[55] and birth defects.[56] It also disrupts digestion by killing the probiotics (friendly bacteria) in the body. Aside from many of the problems inherent in municipal water, proper carbon, sediment and fluoride filtration can bring most of it up to acceptable drinking standards.

Giving unfiltered municipal water to an infant is dangerous because anything that is put into an infant's body is magnified many times because of its tiny size. Bathing an infant in chlorinated water is also harmful, as it is for anyone.

Showering in unfiltered municipal water can be quite dangerous since the body absorbs more chlorine through the skin pores and lungs during a hot shower than it would if you were to drink an eight ounce glass of the same water. Skin pores absorb substances such as chlorine up to twelve times the normal rate in a hot shower because they are fully dilated. I strongly recommend a KDF shower filter to remove chlorine and other contaminants from municipal water. I personally do not travel anywhere without one.

All water ionizers have a built in carbon filter that will remove both chlorine and chloramines.

Fluoride and Its Dangers

"'State', I call it, where they all drink poison, the good and the wicked."

~ Friedrich Nietzsche

Sodium silicofluoride, or fluoride for short, is injected into most municipal water supplies on the premise that it reduces cavities in children's teeth. Fluoride is derived from the processing of nitrogen fertilizer. It can be effectively removed with an activated alum filter, which uses a natural form of aluminum.[57] Fluoride is also present in many processed foods such as soft drinks (1 part per million (ppm)), milk (3 – 9 ppm), breakfast cereal (5 – 15 ppm), and juice (3 – 7 ppm), however these are in the form of calcium-fluoride, which is not harmful. Municipal water contains an average of .6 – .9 ppm fluoride. High levels of fluoride (over 4 ppm) can lead to dental fluorosis or pitting of the teeth. It can also adversely affect the central nervous system. Below are quotes on the practice of adding fluoride to drinking water by researchers, scientists and others in the medical community.

"I am appalled at the prospect of using water as a vehicle for drugs. Fluoride is a corrosive poison that will produce serious effects on a long-range basis. Any attempt to use water this way is deplorable."[58]

"Based on data from the National Academy of Sciences, current levels of fluoride exposure in drinking water may cause arthritis in a substantial portion of the population long before they reach old age."[59]

"Fluorides are general protoplasmic poisons, probably because of their capacity to modify the metabolism of cells by changing the permeability of the cell membrane and by inhibiting certain enzyme systems."[60]

"The plain fact that fluorine is an insidious poison, harmful, toxic and cumulative in its effects, even when ingested in minimal amounts, will remain unchanged no matter how many times it will be repeated in print that fluoridation of the water supply is safe."[61]

The preponderance of evidence that demonstrates what a poison sodium fluoride actually is calls into question the wisdom of its continued use in municipal water systems.

The Center for Disease Control cites fluoridation as one of the ten greatest achievements in health of the 20th Century: *"Fluoridation safely and inexpensively benefits both children and adults by effectively preventing tooth decay, regardless of socioeconomic status or access to care."*

And because of their recommendation many municipalities decide to add fluoride to their water. Fluoridation is also strongly promoted and encouraged by the American Dental Association. In fact, they are one of the driving forces behind the fluoridation of municipal water in the U.S.

Fluoride levels are suggested, not mandated, by the EPA (Environmental Protection Agency). And since fluoride is not a drug, it cannot be regulated by the FDA. The reason municipalities fluoridate their water supply is because they are often pressured by state regulators to use it on the basis that it is less expensive to prevent cavities in children with fluoride in the water supply than it is to subject each of them to fluoride dental treatments. In fact, it's only pennies to use fluoride in the water supply to achieve the same results as spending \$5 – \$10 per child using the yearly treatment method. That's if you accept the fact that adding fluoride to water prevents dental decay in children. Regardless, the question that remains is whether it is worth it to sacrifice our health for our teeth.

The biggest problem with adding fluoride to municipal water is that it interferes with the absorption of iodine, one of the most important minerals required by the body. Without iodine, the thyroid cannot produce thyroxine and triiodothyronine. These hormones regulate growth, protein synthesis, metabolism and a host of other bodily functions. The thyroid is the master gland in the body in that it helps regulate many of other glandular functions and hormone production. The cascading effect of hormonal imbalances leads to disastrous health consequences. Fluoride, chlorine, bromine and iodine belong to group 17 on the periodic table. All have the same valence (1, 3, 5, 7), except fluoride has a valence of 1 only, which makes it the easiest for the body to absorb. Since fluoride, chlorine and bromine are all considerably smaller and lighter than iodine they are also much easier for the body's iodine receptor sites to bond with. These other competing halides make iodine absorption quite difficult, which leads to tremendous imbalances in the body including weight gain, metabolism problems and lethargy.

Halogens (Group 7) in the Periodic Table

1	2											3	4	5	6	7	0
							H										He
Li	Be											B	C	N	O	F	Ne
Na	Mg											Al	Si	P	S	Cl	Ar
K	Ca	Sc	Ti	V	Cr	Mn	Fe	Co	Ni	Cu	Zn	Ga	Ge	As	Se	Br	Kr
Rb	Sr	Y	Zr	Nb	Mo	Tc	Ru	Rh	Pd	Ag	Cd	In	Sn	Sb	Te	I	Xe
Cs	Ba	La	Hf	Ta	W	Re	Os	Ir	Pt	Au	Hg	Tl	Pb	Bi	Po	At	Rn
Fr	Ra	Ac	Rf	Db	Sg	Bh	Hs	Mt	Ds	Rg							

Halogens Weights and Form

Halogen	Molecule	Weight	Size
Fluorine	F_2	18.9	
Chlorine	Cl_2	35.4	
Bromine	Br_2	79.9	
Iodine	I_2	126.9	

Common table salt is iodized, meaning that it has had iodine added to it as nutritional supplement. Ironically, the chloride in sodium-chloride, or table salt, reacts more easily with iodine receptors. The use of chlorine to disinfect municipal drinking water and swimming pools also interferes with iodine uptake. Therefore, drinking or showering in municipal water is strongly discouraged.

"The clinical activity of any one of these four halogens is in inverse proportion to its atomic weight. This means that any one of the four can displace the element with a higher atomic weight, but cannot displace an element with a lower atomic weight. For example, flourine can displace chlorine, bromine and iodine because flourine has a lower atomic weight than the other three."[62]

I understand the logic behind the decision to fluoridate municipal water, although I maintain that the practice of adding fluoride to municipal water supplies should cease immediately. However, we should also acknowledge that we live in a sea of toxins, both natural and synthetic. Heavy metals, petroleum-based elements and a myriad of other poisons have been littering our environment since industrialization began hundreds of years ago. The simple act of riding behind a bus exposes you to a tremendous amount of carbon dioxide, ash, soot, benzene and a host of other toxic chemicals, many times more than drinking gallons of water that contains sodium fluoride. So while I condemn the practice of adding fluoride to municipal water, I like to put into perspective the amount of other toxins in our immediate environment that are far more dangerous. How we remove these toxins from the body should be our greatest concern. One key component for doing so is with the daily consumption of Ionized Water.

Ionized Water and Fluoride Removal

Some have claimed that water ionizers remove 100% of all the fluoride in the source water by sending it all down the drain through the acid water tube because fluoride is an acid mineral. Complete fluoride removal through ionization is a physical impossibility since 100% of the minerals would need to be ionized to remove 100% of any mineral either to the alkaline or acid water tubes. This would mean that the pH of the *Alkaline Ionized Water* would be 14 and the pH of the *Acid Ionized Water* would be 0. Tests demonstrate that the removal rate of fluoride through the *Acid Ionized Water* side is approximately 45 – 50% on the strongest ionization level (pH 10), meaning that about 50% of the fluoride is removed from the drinking side.

Well Water

Many people are under the false assumption that nearly all ground, surface and spring water is contaminated. Most ground water drawn below 150 feet is suitable for drinking if properly filtered, especially in areas that have a lot of clay since it serves as a natural barrier. Clay forms a natural barrier that doesn't allow contaminants to seep into the geological strata below it where water is drawn. In areas that have been chemically contaminated, such pollutants have never been found below the clay strata. Agricultural chemical run-off does not affect ground water below 75 feet if there is clay layer in the strata.

If a septic system is within 75 feet of a drinking well, a bacteria test should be run on the water to be certain there is no cross-contamination from it. High iron content in the source water can cause problems for a water ionizer, although iron can be removed with a carbon or iron resin filter. Pre-filters can be used on the countertop or installed directly into the water line below the sink. There are whole house iron filters that use various media such as green sand, which are costly, but effective. High iron content in well water supplies is more often caused by the well casing and corroded pipes than actual high levels of iron in the ground water.

If the water supply smells like rotten eggs, it contains hydrogen sulfide, leading many to believe the water is undrinkable. Hydrogen sulfide is not harmful and can be filtered out using a KDF 85 filter followed by a granulated carbon filter. Boiling water also effectively removes hydrogen sulfide.

Unless you have had a specific test conducted to determine if you have contaminants in your well water, such as herbicides, pesticides or radon, then chances are the ground water you have is not contaminated.

Softened Water

"What's one man's poison, signor, is another's meat or drink.

~ Beaumont and Fletcher

Never drink water than has been run through a salt-water softener. Salt-softened water should also never be used with a water ionizer, in part because it will overheat the unit due to salt's exceptional conductivity. Sodium is the medium that conducts electrical activity in the body. Drinking salt-softened water throws the body's metabolism out of balance and sends its blood pressure soaring. The salt sold for water softeners is poor quality and should not be consumed. Salt is also corrosive to the water pipes in a house.

Potassium water softeners are a slightly better choice, but given the quality of the potassium used and the amount that would be consumed, I discourage the consumption of potassium-softened water as well.

The best strategy for addressing water hardness is a magnetic water conditioner which provides many of the effects of softened water without the need for salt. Rather than corrode pipes, it actually helps keep them clear, as well as extends the life of your hot water heater by keeping it clear of mineral-scaling. Corrosive salts have the opposite effect that result in the accelerated decay of a hot water heater.

Magnetic water conditioners polarize the minerals in the water so they are all the same polarity, which causes them to line up with one another. The result of this is that they won't stick to any surfaces. As the water comes in contact with the minerals that have previously scaled the interior of the pipe, they also become polarized and are released into the water, which descales the interior of the pipe. This descaling will continue for weeks, months and even years. Evidence of mineral polarization with a magnetic water conditioner is demonstrated by a lower surface tension, which is measured in dynes (dyn). The internal descaling of pipes is further evidence of mineral polarity.

Chapter 4

The Characteristics of Ionized Water

"Life is a struggle, not against sin, not against money or power, not against malicious animal magnetism, but against hydrogen ions"

~ H.L. Mencken

***Ionized Water* is known by many names:** *Alkali Water, Alkaline Water, Alkalized Water, Cluster Water, Microcluster Water, Reduced Water, Miracle Water, Micro Water, Ion Water, Ionic Water, Electron Water, Hydroxyl Water, Electrolyzed Water.*

Alkaline Ionized Water is by far the most superior drinking water available. *Ionized Water* is electronically enhanced water created through electrolysis. It is produced by running normal tap water over negative (cathode) and positive (anode) electrodes, which ionizes the minerals in the water creating positive (hydrogen) and negative (hydroxyl) ions. The electrodes are composed of titanium, the hardest metal known, and coated with platinum, which is an excellent and durable conductor. It is important to note that the platinum is not electroplated, but coated onto the titanium, meaning the titanium plate is dipped into the platinum. Electroplated platinum would not last long on the surface of the titanium. The membranes between the electrodes are composed of either pulp mesh or complex plastic polymers. Polymer membranes are more likely to warp if exposed to high temperatures. Pulp mesh holds up to high temperatures much better than plastic polymers.

The magic comes when the membranes separate the hydrogen and hydroxyl ions, thus creating alkaline and acidic water. These two waters are always produced simultaneously during the ionizing process, 70% *Alkaline Ionized Water* and 30% *Acid Ionized Water*. Therefore, producing one gallon of *Ionized*

Water yields approximately 0.7 gallons of *Alkaline Water* and 0.3 gallons of *Acid Water*.

Electrolysis is most often conducted by placing the anode and cathode in the same solution without a membrane barrier between them. The negative and positive ions cancel one another out, usually causing strong discoloration of the water. The semi-permeable membrane that separates positive and negative ions into alkaline and acid water is one of the great health breakthroughs of the 20th Century.

The properties of *Alkaline Ionized Water* are that it is an antioxidant, alkalizing, hydrating and detoxifying.

The Irony of *Ionized Water*

The production of *Ionized Water* essentially turns anabolic/catabolic processes on their respective heads. The most powerful liquid antioxidant known as *Alkaline Ionized Water* is produced using a cathode and not an anode as one would suspect. In the case of *Ionized Water* we produce alkaline anabolic water filled with negatively charged ions with the very thing we must avoid if we wish to be healthy, that of a cathode, which is associated with catabolic processes. When something is decaying, it means that it is catabolizing or wasting away. The anode, something we normally associate as being good for us (anabolic), produces acid catabolic water, which is harmful if consumed.

Ionized Water: A Meaningless Term?

"Still a doubter? I bought a water ionizer and hooked it up to my kitchen faucet. Our son, Dalton, has struggled for years with asthma. No doctor could help him. Once the water ionizer was installed and he started drinking from it, his asthma improved. Not bad, wouldn't you say?"

~ Bill Romanowski[63]

Some have claimed that *Ionized Water* is a meaningless term because the ionization of water is impossible. The ionization process does not ionize water, but rather the minerals in the water, which results in the production of hydroxyl (-) and hydrogen (+) ions. Technically speaking, *Ionized Water* only describes what we are consuming, which is water that has an abundance of electrons in it. The more minerals in the source water, the stronger the ionization will be, which translates into a higher pH and lower ORP on the alkaline side, and a lower pH on the acid side. *Ionized Water* does not exactly describe what has happened to the water, but it is an easy way for the average person to understand what they are drinking.

Therefore, *Ionized Water* is a term of convenience. More accurately it should be called *ionized-mineral water.* However, *Ionized Water* is a suitable term that is short-hand for what is actually being produced and consumed, which are ionized minerals and hydroxyl ions contained in water.

What is Ionization?

"A vital body equals having spare electrons."

~ Annie and Dr. David Jubb

To *Ionize* means to gain or lose an electron. Essentially, the ionization process robs an electron from one molecule and donates, or transfers, it to another molecule. In the case of *Ionized Water*, an electron is grabbed from one oxygen molecule and donated to another oxygen molecule. This becomes a hydroxyl ion, which is a molecule that carries an extra electron (OH-). Although they are transitory, molecules with this charge produced by the presence of an electron are more easily absorbed by the body.

The presence of hydroxyl ions makes the water alkaline, meaning it has a high pH. An excess of negatively charged hydroxyl ions results in a pH above 7.0. A deficit of electrons results in a low pH below 7.0. An equal amount of electrons in a solution or in the body results in a balanced pH of 7.0.

The other water produced during the ionization process contains molecules that have been robbed of an electron. These are known as hydrogen ions (H+) and they are what make the water acidic, resulting in a low pH (See *Acid Ionized Water* Section).

These yin/yang waters produced by ionization are the exact opposite from one another. Both *Alkaline* and *Acid Ionized Water* have extraordinary properties and benefits, although their respective uses could not be more different. We consume the *Alkaline Ionized Water*. The *Acid Ionized Water* should never be consumed. *Ionized Water has a beneficial effect on everything it comes in contact with as long as it is used properly.*

Ionized Water boils and cools approximately 20 – 25% faster than conventional water due to its smaller molecule cluster size. *Ionized Water* produces a kind of inexplicable buoyancy in a person immediately after consuming a tall glass of it at pH 9.5 or higher. If nothing else, ionization demonstrates the unusual nature of water that has been separated into its basic ionic components that are natural opposites from one another.

Ionized Water is one of the most significant preventative health advances of our generation because it is the most beneficial substance available to the human body. As far as importance, the invention of the water ionizer ranks with the great achievements of the 20th Century along with the man first walking on the moon and the advent of the personal computer.

Ionized Water is an **Antioxidant** that provides the body with an abundance of oxygen, which gives us energy. It possesses a negative charge, or ORP, which is also an antioxidant. It balances the body's pH, which helps prevent disease because it is **Alkaline**. It is a **Powerful Detoxifier** and **Superior Hydrator** because of its small water molecule cluster size.

Powerful Antioxidant

"Accept a miracle."

~ Edward Young[64]

The showpiece of *Ionized Water* are its antioxidant properties. It is truly miraculous that normal tap water can be instantly transformed into a strong antioxidant. *Ionized Water* has two antioxidant qualities, its negative charge and the presence of hydroxyl ions.

Water has a low atomic weight (18[65]) and when ionized it becomes the most absorbable antioxidant known. Other antioxidants have a much higher molecular weight, which make them less easily absorbed by the body.

All liquids have an Oxidation Reduction Potential (ORP), which is the millivoltage (mV), or vibration, it possesses. Vibration is a measurement of frequency, which can be quantified with the use of a frequency meter.

It is said that water has a memory, meaning that it always retains the same ORP unless an external force such as distillation, ionization or other reactive forces change it. The ORP of the water is what water "remembers" because ORP determines the size and shape of water molecule clusters, as well as its surface tension. Normal tap water has an ORP of +300 to +400 mV. Its potential for reducing oxidation is nonexistent because its ORP is above zero. Any number over zero indicates that the potential for increased oxidation is present. The higher the ORP, the more oxidation potential the substance possesses. This means that if it comes in contact with another substance it has a greater potential to oxidize or cause that substance to decay. The ORP environment we create within ourselves directly reflects the state of our health.

Accelerated Aging vs. Reverse Aging

"[Ionized Water] is one of the simplest and most powerful things you can do to combat a wide range of disease processes."

~ Ray Kurzweil[66]

A negative ORP can reduce, or negate, oxidation. Strong *Alkaline Ionized Water* has an ORP of -50 mV to -450 mV, depending on the source water and how many minerals it contains. The more minerals in the source water, the stronger the *Ionized Water* will be produced. This low negative number means that the water has a very high potential for reducing oxidation. A beverage that has an ORP of -350 mV is healthier to consume than -150 mV because it negates oxidation of the body more effectively. Therefore, the lower the ORP of the water, the greater potential it has to reverse the aging process of the body at a cellular level.

Fresh squeezed raw orange juice has an ORP of about -250 mV. All fresh squeezed vegetables and fruit juices and vegetables have a negative ORP, some lower than others. Therefore, they are considered antioxidants because they reduce the potential for oxidation in the body. However, if these juices are heated above 118 F°, pasteurized or otherwise processed, the negative ORP antioxidant property is destroyed. In fact, all its rejuvenation properties have been removed and now the food has been transformed into mere sustenance that provides the body with calories, almost no nutrition and helps to acidify it. Enzymes must be present in a food for it to truly be considered a rejuvenating substance. This same principle is true for *Ionized Water*. If it is heated, it will quickly lose its negative charge because the fragile, fleeting electrons will be destroyed. Electrons are thousands of times lighter than protons, thus they are more easily dispersed and scattered than protons.[67] However, *Ionized Water's* other properties such as alkalinity and reduced water-molecule cluster size remain intact to some degree for a longer period of time.

As substances become oxidized, their ORP rises. Oxidation means to react with oxygen. Rust is metal that has been oxidized, which is an example of slow oxidization. Fire is an example of fast oxidation. In the human body, oxidation is caused, in part, by free radical damage. Unstable oxygen molecules rob us of electrons, which causes oxidation, leading to aging and disease. *"One can thrive on half the normal intake of food as long as we consume high electron-rich nutrients."*[68] Any time we put a substance in the body that has a positive charge, we increase the oxidation of the body and therefore accelerate the aging process. As we age, our body's ORP continually rises. The pace at which our body oxidizes is directly related to our diet and the other substances we put in it. Our immediate environment also contributes to the oxidation of the body. Genetics does not determine the rate of oxidation of the body.

A polluted environment will increase the oxidation of the body by creating free radicals in it. Pollution contains an abundance of positive ions that is detrimental to our health. We breathe in 11,000 liters of air each day, thus the quality of the air we breathe is of great importance. When we breathe in polluted air, toxins are pumped directly into our bloodstream. Air purifiers, such as ozone generators, negative ion generators and air filters are imperative if we wish to breathe clean air and remain in a healthy environment.

Alkaline Ionized Water has a negative ORP, therefore it offsets the positive ORP of our oxidizing, aging body. Thus, we counteract the aging process by consuming negatively charged substances, which dampen the positive ORP of our oxidizing body. Realistically, we need to drink at least 1 to 2 gallons of strong *Alkaline Ionized Water* each day if we expect significant slowing and reversal of the biological aging process, which is determined by the health of our cells. Human health equals cellular health. If our body's cells are not healthy nor can we be healthy.

Charge of Life

"Photons from the sun energize electrons in our body through resonance."

~ Annie and Dr. David Jubb

Consuming fresh *Ionized Water* puts an electrical charge into the body the same way that raw fruits and vegetables do. Raw foods have this charge because they are full of enzymes, electrons and electrical activity, which is facilitated by the mineral content of the plants the same way it is for us. Raw foods are bio-photonic, meaning they are created by sunlight (photons) and they are alive with enzymatic activity. Raw foods are essentially concentrated sunlight. Thus they are bio-electrical, meaning they are alive with electrical activity. We are also bio-electrical, meaning that electricity conducts through the body when there are sufficient amounts and varieties of minerals present for electricity to flow. If these minerals are absent from the body it will not function properly. In the total absence of these minerals, the body and its organs will cease to function and we will die.

Strong, fresh *Ionized Water* charges the body because it contains large amounts of electrons that encourage electrical activity in the body. *"The more alive something is, the more it is moving from the dense matter of nucleons and protons to the world of light and electrons."*[69] The foods and water that we consume should contain substances that promote electrical activity because they provide us with energy and charge the body's internal battery. *Ionized Water* also has a negative charge, or ORP, that promotes electrical activity in the body. When the body is charged and has sufficient amounts of electrical activity we feel energetic. The brain cannot operate without electrical activity. Every thought we have produces a miniature electrical storm in various areas of the brain, depending on what the thought is. As we stimulate electrical activity in the brain by consuming substances that encourage electrical activity, we are able to think more clearly. It is this constant recharging of the body through raw foods and *Ionized Water* that helps keep us young, active and disease-free.

The substance that puts a charge in the body, does so in the form of a small amount of energy measured in millivolts. Energy provides the body with vitality and clear thinking. It doesn't make us want to bounce off the walls or sprint five blocks. Thus, we should not confuse the two. Energy provides the body with greater potential for health and vibrancy. Stimulants such as caffeine and other additives do nothing but jolt our nervous system then leave us flat and burned out once they have been used up.

The Most Important Term in Health

"You're never too old to become younger."

~ Mae West

ORP is the single most important term we need to become familiar with if we want to understand human health. A person's ORP level, although quite difficult to determine reliably, would instantly reveal whether they are in a state of health or disease. ORP is another way to measure the body's vibration. Everything in the universe vibrates. When we are healthy we vibrate within a certain frequency range. If we are sick, we will vibrate at a completely different frequency range, one that reflects our state of unhcalthiness or disease.

At 49, I personally am not in physical decline, in large part because I regularly drink *Ionized Water*. What the rest of my diet is comprised of also determines my current biological age quotient. The cells that make up my body are more active, productive, communicative and functional than those I had when I was supposedly in my prime 30 years ago, although given my health warrior lifestyle I'm definitely in my prime now. The great thing is that 10 years from now I'll still be in my prime and look back at how much healthier I am then compared to today.

ORP is a measurement of a substance's ability to either reduce or encourage the oxidation of another substance. When we consume raw foods, they reduce the oxidation of our bodies. Thus, raw foods rejuvenate us. Raw foods are also negatively charged. Cooking raw foods oxidizes them, thus raising their ORP. And when we consume cooked foods, we add to the oxidation of our bodies and accelerate the aging process. Cooked foods burn us up internally by stimulating oxidation since they themselves have already been oxidized with a positive ORP of +400 or higher. Animal protein, fried foods, soft drinks and other highly processed foods possess the highest ORP and therefore greatest amount of hydrogen (positive) ions. A high ORP is an *environment* where disease thrives because it is also a high acid (low pH) *environment*. To reduce this oxidation, this slow-burning fire within us, we must consume substances that possess a negative charge such as *Ionized Water* and raw fruits and vegetables. When we do, the consuming fire of high ORP is extinguished.

The principals of ORP are the same for *Ionized Water*. The positive ORP of *Acid Ionized Water* increases oxidation because it contains hydrogen ions (missing electrons), which is the *environment* of all disease. The negative charge of *Alkaline*

Ionized Water reduces oxidation because it contains hydroxyl ions (extra electrons), which is an *environment* that leads to health.

Consuming *Ionized Water* bathes the interior of the body in a negatively charged liquid, which promotes rejuvenation of each bodily system at a cellular level. For instance, a liver cell is better able to repair itself in a negative ion, alkaline environment than a positive ion, acid environment. When we consume negatively charged substances such as *Ionized Water* this oxidation is retarded and our body's cells are in a better position to repair and rejuvenate themselves. Nothing is better for the body.

How the Body Decomposes Before We Have Died

They burn, that is, exhale and decompose their own bodies into the air and earth again.

~ Ralph Waldo Emerson

Hydrogen ions cause oxidation and decay. Hydrogen ions in the body are referred to as free radicals. They can damage the DNA of a cell when they steal an electron from it. If the cell does not die, it will reproduce using a corrupted set of DNA instructions. Both cells will be slightly mutated as if a few critical pages of an instruction manual had been altered or entirely removed. These mutated cells with the now incomplete and/or damaged DNA instruction set are the cause of many diseases for obvious reasons. If cells are not replicated with the correct human genetic code, the body now has cells in it that are foreign to its natural state, which is a detriment to the body as a whole. Negating hydrogen ions with hydroxyl ions in the body will retard the onset of disease and has the potential to reverse the biological aging process itself.

"Because active oxygen can damage normal tissue, it is essential to scavenge this active oxygen from the body before it can cause disintegration of healthy tissue. If we can find an effective method to block the oxidation of healthy tissue by active oxygen, then we can attempt to prevent disease." [70] This is accomplished by drinking *Ionized Water*.

Drink the Cloud: Increased Oxygen

"Electrons have an affinity for oxygen."
~ Annie and Dr. David Jubb

A fresh glass of strong *Alkaline Ionized Water* right out of the tap will contain a cloud of tiny bubbles in the water. These are hydroxyl ions, *Ionized Water's* other antioxidant component. The best way to drink *Ionized Water* is as fresh as possible. Drinking cloudy *Ionized Water* with its abundance of electrons is one of the healthiest things we can do.

"Oxygen is essential to survival. It is relatively stable in the air, but when too much is absorbed into the body it can become active and unstable and has a tendency to attach itself to any biological molecule, including molecules of healthy cells. The chemical activity of these free radicals is due to one or more pairs of unpaired electrons. Such free radicals with unpaired electrons are unstable and have a high oxidation potential, which means they are capable of stealing electrons from other cells." [71]

Some antioxidants possess an extra electron. Others, such as carotenoids, which are natural pigments from foods, retard and reverse aging through chemical processes. *Ionized Water* is an extremely effective antioxidant because it is a liquid with small water molecule clusters, and it is more easily absorbed into the body where it can be of immediate use.

"When taken internally, the reduced Ionized Water with its redox potential, or ORP, of -250 to -350 mV readily donates its electrons to oddball oxygen radicals and blocks the interaction of the active oxygen with normal molecules." [72]

"Alkaline water has a high negative redox potential which quickly permeates the cells and its abundant electrons to free radical oxygen molecules. This produces an excess of electrons to donate to peroxide and hydroxyl radicals." [73]

Antioxidants have anti-aging and anti-disease properties because they help return the body's cells to a youthful, healthier, more natural state. As we make *Ionized Water* a part of our daily routine and drink sufficient quantities of it, we begin to bathe the body's cells in alkalinity and antioxidants while at the same time hydrating them better than they have ever been. Nothing could be more fundamentally healthier for us.

"Problems arise, however, when too many of these active oxygen molecules, or free radicals, are produced in the body. They are extremely reactive and can also attach themselves to normal, healthy cells and damage them genetically. These active oxygen radicals steal electrons from normal, healthy biological molecules. This electron theft by active oxygen oxidizes tissue and can cause disease." [74]

"There is no substitute for a healthy balanced diet, especially rich in antioxidant materials such as vitamin C, vitamin E, beta-carotene, and other foods that are good for us. However, these substances are not the best source of free electrons that can block the oxidation of healthy tissue by active oxygen. Water treated by electrolysis to increase its reduction potential is the best solution to the problem of providing a safe source of free electrons to block the oxidation of normal tissue by free oxygen radicals. We believe that reduced water, water with an excess of free electrons to donate to active oxygen, is the best solution . . ." [75]

Free radicals are another example of the environment that encourages disease in our body by causing cell mutations and other types of cellular damage. Free radical cellular damage is a big part of the aging equation, but it also can be reversed with proper diet and the consistent use of *Ionized Water*.

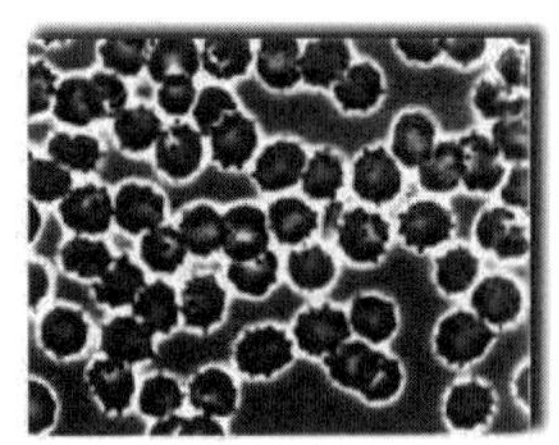

Free Radicals are commonly created from:

- Prescription and street drugs.
- Chemicals (pesticides, herbicides, insecticides, etc.).
- Processed and irradiated foods.
- Food additives and preservatives.
- Heavy metal poisoning.
- Artificial food colorings.

- Polyunsaturated oil, mainly vegetable oils, and rancid oils.
- Trans-fats (partially hydrogenated fats).[76]
- Chlorinated unfiltered tap water.
- Tobacco use.
- Excessive, prolonged stress.
- Cooked foods of all kinds, especially fried foods.

Poikilocytes are an example of free radical damage that has occurred to blood cells. Poikilocytosis is an increase in the number of abnormally shaped red blood cells that is an indication of oxidation of the cell.

Medical Science: Endlessly Searching Down Dead End Roads

Essentially all disease appears and develops in the body from our diet, unless it is congenital in nature, meaning that it is a condition that exists at birth. However, this almost always has to do with the diet and environment of the host mother of the child that is born. Poor diet, dehydration and a polluted physical environment bring about most congenitally caused diseases and birth defects such as Down's Syndrome, mental retardation and premature birth. The roots of their origins do not lay in genetics. Science can determine which genes have been damaged or otherwise altered, but it has only theories as to why and how the defect actually occurred. Environmental diseases may come from sources such as artificial chemical toxins, heavy metals, radiation, naturally occurring toxic substances, or insect-borne diseases, but these can all be overcome by the right diet, *Ionized Water*, probiotics, exercise and use of a FIR (Far Infrared) sauna.

The scientific and medical communities are desperately trying to find a genetic link to every disease, including cancer. It is an effort to establish that disease is born in genetic anomalies and flaws, thus curing these diseases lie in the engineering and reconstruction of these flawed genes. These efforts manifest themselves in medications and artificial therapies of every kind. The latest effort is vaccinations for diseases such as the vaccination developed for cervical cancer.[77] All these efforts are enabling devices designed to allow us to keep eating foods that we love and are accustomed to, but unfortunately lead to all disease. The path to health does not lie in these artificial protocols and procedures, nor will it ever. True health is found only in nature.

Oxygen and the Antioxidant Connection

Drinking Ionized Water gives you energy. On the surface, it seems like an outrageous claim that drinking water could possibly give you energy. However, strong, fresh *Ionized Water* possesses an abundance of hydroxyl ions, which destroy unstable free radicals by donating an electron to them. What is left behind are stable oxygen molecules, which provides the body with more dissolved oxygen. If your blood oxygen level is low, check it before you first start drinking *Ionized Water* and then again a few weeks after you have been drinking it regularly and see the difference for yourself.

Energy derived from oxygen is the best kind of energy because it has not been derived from a source such as caffeine, sugar or chemical stimulants. These substances produce an artificial rush that feels like energy, but eventually lets us down once they are used up. For instance, when we consume too much sugar, the pancreas produces insulin to counteract the sudden influx of sugar to the bloodstream and we experience a sugar/insulin crash. Years of this kind of activity can lead to diabetes, one of several diseases that has become epidemic amongst children because of their high sugar diet. Offering a child sugar-laden cereals and processed juices for breakfast on an empty stomach invites trouble in the classroom that will eventually manifest itself as hyperactivity and disease of every kind, most commonly diabetes and obesity.[78]

With oxygen, it's an entirely different story. The body never gets a rush from oxygen and never crashes from it because it is naturally derived energy. People who drink *Ionized Water* on a regular basis often don't notice how much energy it gives them until the fourth or fifth day after they stop drinking it and the oxygen-induced energy has finally dissipated.

Stable oxygen is a nutrient the body desperately requires. It provides us with mental alertness and invigorates the body. It also carries vital nutrients around the body. However, the body cannot store oxygen the way it can with other nutrients such as copper or iron. When oxygen becomes saturated in the blood, the body quickly discards any surplus before it becomes unstable and causes oxidation. Oxygen is constantly used by the body in numerous capacities. Therefore, we must constantly provide the body with more oxygen and there is no better natural source than *Ionized Water*.

For someone with cancer, the high oxygen level provided by *Ionized Water* is particularly helpful since oxygen reportedly destroys cancer cells.[79] Cancer patients are often tired because their blood oxygen levels are low after being depleted by the process of mutual destruction between cancer cells and oxygen molecules. Drinking fresh, strong *Ionized Water* on a regular basis will increase the amount of dissolved oxygen in the bloodstream. This is helpful in fighting many diseases since oxygen thoroughly cleanses the blood. Therefore, the more stable oxygen we have in our bloodstream, the better our health will be. Stable oxygen in the bloodstream means that the oxygen molecule has evenly paired electrons and therefore is non-reactive. Reactive oxygen is commonly referred to as free radicals.

Ionized Water and the Death of Bacteria

Bacteria do not live in strong *Alkaline Ionized Water* (pH 9.9+) much more than 40 seconds because it is an oxygen-rich alkaline environment. It has a low negative charge (-200 mV to -400 mV), which is also hostile to bacteria. Bacteria are killed instantly in *Acid Ionized Water* because of the high charge in the water (+900 mV). At +600 mV (+/-) bacteria begin to die. +1100 mV is considered anti-microbial, meaning that bacteria absolutely cannot survive in such a highly charged environment because it literally electrocutes them. Bacteria thrive best in a mV range near zero, not the extreme pH and ORP environments produced by ionization.

Alkalizes and Balances Body pH

"That which is built on alkalinity sustains: That which is built on acidity falls away -- be it civilizations, human bodies, or the paper that preserves their knowledge."

~ Dr. T. Baroody

The world's written history was recorded on alkaline paper until 1850 when it began to be recorded on paper that used bleach, alum and tannin in the book-binding, all of which are acid. Those original written records starting in 1850 are disintegrating at an alarming rate. The best that can be done is to scan them electronically and save what is left of the books by re-alkalizing the remaining paper. However, books printed on alkaline paper before 1850 still survive, often in perfect condition. Acid destroys life. A balanced, slightly alkaline pH, preserves it.

We look everywhere for health when it is never any further than what we put in our body. Alkaline substances belong in the body, not acid ones. *All disease thrives in an acid environment and will not thrive in an alkaline environment.*

Dramatically lowering the pH of the body causes many enzymatic reactions to cease, putting cellular metabolism itself at risk. Enzymes and enzyme reactions are some of the most basic and important functions of the body. In fact, if enzyme activity were to cease, life itself would as well. Therefore, it is imperative that we maintain a proper pH and mV range so that enzymatic activity functions optimally.

As the acid condition of the body worsens, it destroys cell walls, corroding veins and arteries and eventually entire organs. The skin of an overly acidic person is markedly more wrinkled, worn, unhealthy and prone to disease compared to someone whose body is more alkaline. The ideal body pH is around 6.8 – 7.0, although you will find that this range is debated to some degree amongst naturalists. I have not found substantial evidence that demonstrates the human body functions best at any other pH than one close to neutral, 7.0.

"Alkaline water, having a pH of between 9 and 11, will neutralize harmful stored acid wastes, and if you consume it every day, will gently remove them from your body. Yet, since the water is ionized, it will not leach out valuable minerals like calcium, magnesium, potassium, or sodium." [80]

If we create an acidic environment in the body by years of consuming cooked foods and other acidic substances such as drugs, alcohol, cigarettes, soft drinks, processed sugar, etc., then we become vulnerable to any disease that invades the body, regardless of its source. The more acidic a person is, the more susceptible they are to disease. This acid environment does not cause disease, but rather creates an environment that disease thrives in. As disease flourishes in the body, it creates a more acidic environment in order to spread further until it consumes the body. Disease is essentially another mechanism that nature uses to recycle something that is no longer a part of itself. Yeast, fungus and mold found in the body are also recycling agents used by nature to depose of that which is no longer belongs to itself. All disease is unnatural and is an outgrowth of the wrong diet and has nothing to do with genetics as we are led to believe. Even if we do have genetic propensities toward certain disease, those diseases can be prevented and overcome by proper diet and hydration of the body. Chronic disease of any kind is never found in nature, only bacterial and viral disease. The only instances we find of animals in the wild with chronic disease are those that have been over-exposed to synthetic toxins in polluted areas.

"Living things are extremely sensitive to pH and function best (with certain exceptions, such as certain portions of the digestive tract) when solutions are nearly neutral. Most interior living matter (excluding the cell nucleus) has a pH of about 6.8."[81]

When some cancer patients are near death their bodies begin to produce ammonia as a natural chemical reaction meant to counteract the extremely acidic environment that has been created in their body by the cancer. This is the bad odor that some cancer victims have as they near death.

". . . acid wastes literally attack the joints, tissues, muscles, organs and glands causing minor to major dysfunction."[82] The medical establishment as a whole does not embrace the concept that body pH and disease work hand in hand. In fact, they still are dubious that there is any link at all. The truth is that if you have a cold, flu or any disease you will have a much better chance of conquering it once you have raised the pH of your body by drinking *Ionized Water*. Drinking herbal tea with fresh, raw lemon will often make you feel better when you're sick because lemon is a strong alkaloid, meaning that it creates alkalinity in the body when consumed. Alkaloids contain nitrogen and create alkalinity in the body through chemical reaction. By raising its pH, the body is given its best chance to fight the viruses, bacteria or disease with the defenses it has in its arsenal. The body best achieves this when its internal terrain is altered toward alkaline conditions. All bodily systems, including the immune system, function optimally when the body's pH is balanced near 7.0.

Body pH should not be confused with blood pH. Blood pH is always stable in a range between 7.25 – 7.45. When anything enters it outside that range, the blood immediately pushes it into the surrounding tissue. Because of this, acid waste tends to collect around the organs and in the joints where it encourages organ disease and arthritic conditions of every kind. One of the keys to Great Health is keeping body pH properly balanced and alkaline. Drinking plenty of *Ionized Water* will help achieve that.

Fresh *Ionized Water* provides us with huge amounts of negative hydroxyl ions, which negates the hydrogen ions, or *free radicals,* that accumulate in the body. Acidity is measured by the presence of hydrogen ions. The more hydrogen ions that are present, the more acid the water becomes since pH (*potential for hydrogen*) measures the presence or absence of hydrogen ions. In the purest of terms, pH is a measurement of electrical resistance between negative and positive ions. If the body possesses an abundance of positive hydrogen ions, we are overly acidic. If the body possesses an abundance of negative hydroxyl ions, we are overly alkaline. Ideally, we should be balanced with an equal amount of both of them, giving us a pH of 7.0.

The Importance of Minerals

The ease at which a 9.9 pH is achieved when ionizing water will depend on how slow the water runs through the water ionizer, as well as the water temperature and mineral content of the source water. Tepid water makes stronger *Ionized Water* than cold water because it allows the electrodes to become hotter. The hotter the electrodes, the stronger the ionization will be. More minerals in the source water will produce stronger *Ionized Water* because minerals provide conductivity. Some water ionizers have built-in mineral ports where calcium and other water soluble minerals can be leached into, or added, to the source water as it passes through the filter before it is ionized. Without conductivity, ionization cannot occur. Purified water, either distilled or reverse osmosis, has no mineral content and therefore almost no ionization occurs if this water is used with a water ionizer because of the lack of conductivity that only minerals can provide.

There are both alkaline-forming and acid-forming minerals. Calcium, magnesium, sodium, iron and potassium are alkaline-forming minerals because their ions have a positive charge. Ionization provides them with a negative charge. Calcium is Ca++; Potassium is K+; Magnesium is Mg2++. These minerals will strengthen the alkaline side of the *Ionized Water* by forming hydroxyl ions. Ionization will become stronger on the alkaline side if there is a predominance of alkaline minerals present in the water and stronger on the acid side if there is a predominance of acid minerals.

The reason they are positively charged is because they are stable with a positive ionic charge. When calcium has a negative charge, it means it has an extra electron attached to it. The calcium molecule will be more easily assimilated by the body because of the extra energy the calcium molecule has due to the electron that is attached to it. Minerals that have this charge are considered ionic.

". . . all ingested substances and all situations (physical, emotional, or mental) that affect the body, leave either an alkaline or acid ash residue in the urine."[83]

Sulfur, iodine, chloride, phosphorous, bromine, copper, silicon and fluoride are acid-forming minerals because they have a negative charge. The body uses various minerals, many of which leave behind acid ash when they are used up. For instance, each heartbeat occurs due to magnesium firing. Nerves must fire in precisely the correct sequence to stimulate muscles that contract the chambers of the heart. As the magnesium is used, the acid ash from the reaction must be removed. The body removes ash efficiently when it is healthy, alkaline and its various systems are balanced. When mineral ash is not removed from the body, it accumulates and acidifies it.

Becoming Too Alkaline

"The best and safest thing is to keep a balance in your life, acknowledge the great powers around us and in us. If you can do that, and live that way, you are really a wise man."

Euripides[84]

It is difficult for the body to become over-alkalized, a condition known as *alkalosis*, which is extremely rare and is caused by unnatural imbalances in the body as a result of poor diet.

Nearly everything the average person consumes, including cooked and processed foods, acidifies the body tremendously. Nearly all recreational beverages are acidic, including coffee, black tea, commercial juices, milk, soft drinks and alcohol. Stress adds tremendous amounts of acidity to the body, as does pollution. Industrialization has toxified and acidified our environment since its inception. Toxins have concentrated at the north and south poles because of air currents and prevailing weather patterns. Given the amount of acid that is added to the average person on a daily basis, it would be extremely difficult for anyone to over-alkalize their body. I have consumed 1.5 – 2.0 gallons of *Ionized Water* everyday for 10 years at a pH 9.5 or higher. I live on a 99%

raw food diet, which is alkalizing, and my body pH is always balanced at close to 7.0. I have never measured my body pH and found it to be too alkaline. Over-alkalizing your body will not occur if your approach to health is completely natural. Nature always puts the body into balance when its laws are followed and at the core of homeostasis is a neutral pH. If we wish to determine a person's overall health, the first determination that should be made is their body pH.

Ionized Water and Stomach Acid

A question commonly asked about Ionized Water is how can the alkalinity of Ionized Water survive stomach acid. Digestive enzymes and stomach acid, hydrochloric acid (HCL), are produced by the body as we chew our food. Once we stop eating, the HCL dissipates in the stomach and no more should be produced until the next time we eat. Many people are under the mistaken notion that the pH of the human stomach is always at a pH of 2.5 – 3.0. In fact, when we are not eating the pH of the stomach should be around pH 4.0 – 5.0. Temporarily changing the pH of the stomach to pH 6.0 or higher is not harmful to the body.

If we assume that drinking ionized water will raise the pH of the stomach higher than it should be or that Ionized Water will be neutralized by our stomach acid, then we must also assume that many alkaline foods that we eat will do the same thing. A lot of fruit tends to have a high pH, therefore it is not uncommon for the stomach to have alkaline foods introduced into it. What is important to consider is the overall effect of foods and other substances that we put into the body. If our diet is mainly comprised of acid foods such as fried foods, meat, processed foods, junk foods, fast foods, etc. our body will turn acidic over time. On the other hand, if we put primarily alkaline substances in our body then our body will become more alkaline. All disease thrives in an acid environment in the body and does not thrive in a neutral pH close to 7.0.

Nausea and Ionized Water

Drinking *Ionized Water* with a high pH of 9.0 – 9.9 can sometimes induce a feeling of nausea in the stomach, especially when consumed first thing in the morning. This is not harmful nor is it an indication that you are consuming *Ionized Water* at too high a pH level. If the stomach remains acid with a pH of 5.0 or lower, the sudden introduction of an extremely alkaline substance such as *Ionized Water* can cause a gag reflex in the body. This is a natural reaction because the body views any radical change in pH as something potentially dangerous or poisonous to it, thus its instinct is rid the body of it. Relax, breathe slowly and deeply and the feeling of nausea will likely subside in a short period of time.

Measuring Body pH

Body pH is measured through the saliva or urine. Many things can cause the pH of urine or saliva to quickly rise or fall. For instance, consuming extremely alkaline substances such as asparagus or *Ionized Water* can turn the urine acidic because they effectively remove acid waste from the body. With persistence, however, consuming alkaline substances will turn body's overall pH more alkaline.

Saliva is an accurate measurement of pH as long as you haven't drank or eaten anything for 45 minutes before measuring it. Litmus pH paper is the most accurate measurement of saliva's pH. Color pH test drops (phenol red) are a little more accurate than litmus paper, but both rely on matching color with the eye, which can be somewhat subjective. It's difficult to determine the exact pH of any liquid within a single logarithmic point, meaning the difference between 9 and 10, let alone a tenth of a point.

For accuracy and reliability, a digital pH meter is best. They accurately measure the pH of *Ionized Water*, other liquids that you consume and the pH of the body itself. However, you get what you pay for. A reliable pH meter will cost US$150 - $200.

Litmus paper is not accurate once exposed to the air or if it is not fresh. Phenol red drops only work for urine since it is difficult to get enough saliva to measure it properly. The same is true for digital pH meters, which are accurate, but can only measure urine. The most accurate urine pH reading may be measured first thing in the morning, however, body pH on any given day can be high or low for any number of reasons. For an accurate profile of body pH it is best taken each morning for several days, weeks and then months to get a moving average. Depending on how alkaline your diet is, consistently drinking a sufficient amount of *Ionized Water* each day will cause a trend of raising the body's alkalinity toward neutral.

Powerful Detoxifier and Superior Hydrator

"That is Indisputable."

~ Voltaire

Ionized Water is sometimes called *Cluster Water* or *Microcluster Water* because of its small molecular grouping. Water molecules typically group in clusters of 10 or more. Cluster size is measured with the use of Nuclear Magnetic Resonance.[85] *Ionized Water* molecule clusters are grouped into six water molecules, thus they are *reduced* in size. The *Ionized Water* molecule cluster has changed from an irregular, clumpy shape to a hexagonal shape that penetrates and saturates body tissue much more efficiently than conventional water.

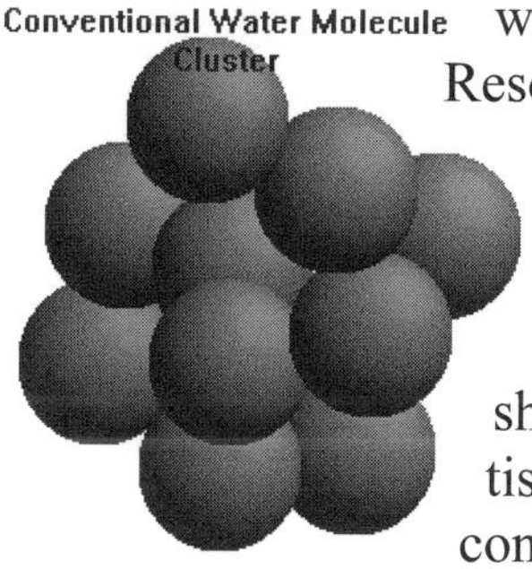

Water ionizers have more than one level of ionization strength, which is important to some people when they first start drinking it. The strong detoxification aspects of *Ionized Water* require that people with a buildup of toxins in their body and tissues begin drinking it at a mild ionization level (pH 8.0), then slowly increase the strength of the *Ionized Water* over the following days and weeks until they acclimate to it. When first using *Ionized Water,* **headaches, rashes, diarrhea and fatigue are common detoxification symptoms** for people who have accumulated toxins throughout their body from poor lifestyle choices. These different levels of ionization strength allow people to slowly ease into *Ionized Water* when they first start drinking it in order to mollify these powerful detoxification effects that can be drastic for those who are quite toxic. The micro-cluster structure and penetrating aspects of *Ionized Water* leave less room for anything that does not belong in bodily tissue. Thus toxins are effectively pushed out of the tissue and into the bloodstream to then be eliminated by the body.

The sight of a glass of *Ionized Water* with small bubbles in it is quite deceiving when it comes to detoxification. In fact, it is a powerhouse when it comes to getting rid of what is not wanted in the body. A toxin is simply something that does not belong in the body.[86] There are mild toxins and quite dangerous toxins such as heavy metals, asbestos or industrial chemical residue. The average person can begin drinking *Ionized Water* without having a serious *healing crisis*, which is a term sometimes applied to a strong and immediate detoxification that is usually an unpleasant, if not painful experience. Those most at risk are people who have taken a lot of street drugs, prescription medications or those whose diets consist primarily of processed, fried, junk or fast foods. Also at risk of a strong detoxification are those who have been exposed to environmental toxins such as heavy metals, herbicides and/or pesticides, which are more common than we realize. This is a tragedy for our health, but we can completely restore our body and our health starting at the cellular level.

As the toxins, or poisons, that disease thrives on leave the body, it undergoes profound changes. Sometimes dependency withdrawals can cause the body to ache and thrash about in agony. Although this can be quite objectionable, you must go through it if you want to be healthy. *Ionized Water* is deceptively powerful because water is conventionally thought to have a neutral effect on the body, not act as a powerful detoxifier.

As long as we don't consume too much in a short period of time or around mealtime, we cannot drink too much *Ionized Water* once the body has acclimated to it. The more toxins a person has accumulated in their flesh, tissue and cells, the weaker the *Ionized Water* should be when they first start drinking it so these unpleasant detoxification effects are kept to a minimum. If the detoxification symptoms become too strong, reduce the strength of the *Ionized Water* and drink less of it. If a person maintains a good diet, drinks a lot of water, doesn't smoke, drink alcohol heavily, take drugs or medication, they can usually start drinking *Ionized Water* at the highest level (pH 9.5 – 9.9).

Regarding children, the vast majority of them have no trouble drinking *Ionized Water* on the strongest level when they first start because they are too young to have accumulated many toxins in their flesh. Their young, resilient bodies are also able to quickly adjust to the healthy environment that *Ionized Water* creates.

The opposite is true regarding the elderly, who have a lifetime of toxins and heavy metals accumulated in their bodies and are often on medications that also need to be detoxified from their tissue. Children on strong medications or a junk food diet may also suffer these powerful detoxification effects as well.

Ionic Detoxification

Toxins are also removed from the body because of the abundance of electrons in *Ionized Water*. Nearly all toxins in the body have a positive charge. When we consume cooked or processed foods, or other toxic substances that have a positive charge, toxins become locked up in the body's tissue and become even more difficult to remove. The negative ions found in *Ionized Water* attach themselves to these toxins and cause them to be released into the bloodstream to then be removed through the liver, kidneys or other excretory organs. *Ionized Water* possesses an abundance of anions that help release and remove cation-laden toxins from the body's tissue.

When consumed, *Ionized Water* moves through the body rather quickly, up to 30% faster than conventional water. *Acid Ionized Water,* however, moves through the body rather slowly if consumed, which is an indication that its effects on the body are the opposite of *Alkaline Ionized Water.* Again, this demonstrates the effects of negative ions (anions) verses positive ions (cations) on the body.

Substances that contain anions such as *Alkaline Ionized Water* move quickly through the body. Substances that contain cations such as *Acid Ionized Water* move slowly through the body because the positive ions react with the negative ions that comprise most of the body. It moves through the body as though it were thick sludge rather than water. Cooked foods, which contain cations, also move slowly through the body. Raw foods, which contain anions, move quickly through the body.

The Clean Temple

"The further you get from nature,
the closer you get to the doctor."

~ Dr. Bernard Jensen

The human body is a temple of God, a more immediate and important one than any church or religious temple. We would not throw trash inside the temple or church where we worship, nor should we put things in our bodily temple that do not belong in it. We consume things that do not belong in the body because we like how they taste and we mistakenly believe that they are good for us. Honor your temple by putting the right substances in it and nature will reward you with Great Health.

Seeing is Believing

"The Facts will appear with the shining of the dawn."

~ Aeschylus

Ionized Water is up to six times more hydrating than conventional water. It is easy to demonstrate the effects of *Ionized Water* regarding its smaller molecule clusters and superior hydration. Soaking items such as dried pinto beans in *Ionized Water* will cause them to hydrate considerably faster and swell fatter than they would after being soaked in conventional water. Mixing *Ionized Water* with any kind of powder or flour will result in much better absorption than using conventional water. Making gravy or cream soup with *Ionized Water* results in a smoother texture and less lumpiness, which is further evidence that the absorption factor of *Ionized Water* regarding any material is greater compared to that of conventional water.

The Environment of Health vs. the Environment of Disease

"They certainly give very strange names to diseases."

~ Plato

If you are sick with any disease, you know two things about your body and its state of health. First, you know that it has a deficit of electrons because your body is acidic. When your body has accumulated too many hydrogen ions (positive cations) then you are acidic and your pH is low. pH is a measurement of the presence of positive and negative ions. When they are present in equal quantities we have reached a state of equilibrium within the body and our pH will be 7.0. An acid environment in the body encourages disease.

If you are sick you also know that you are toxic because disease lives on toxins. Without the presence of toxins, disease cannot exist in the body. If you are healthy your body will not contain substances that do not belong in it. Anything that does not perform a useful function in the body should be regarded as a toxin that needs to leave the potentially glorious palace we call the human body. However, the body is only truly glorious when it is free of toxins. Toxins turn the body into a garbage hull. *Ionized*

Water creates an environment of health by making us more alkaline and removing toxins from the body.

The Perfect Reflection of Nature

"When Nature has work to be done, she creates a genius to do it."

~ Ralph Waldo Emerson

In nature, *Ionized Water* is most often found in mountain streams where water bounces off rocks, which results in a mild ionization effect by the creation of hydroxyl ions. The ionization of water also occurs in other places such as water falls or waves crashing onto a beach. These two natural events also create negative ions in the environment. The hydroxyls that are created under these circumstances are short-lived due to the reactive substances that surround them. They are also cancelled out by positively charged hydrogen cations. We are surrounded by active oxygen in the environment and throughout our diet, which we call free radicals.

"Ionization happens in the body as oxygen and hydrogen atoms gain or lose electrons, making them electrically charged and therefore more chemically active."[87]

A balanced state of health is that in which the body thrives and is most apt to operate to its capacity. However, if you change the body's environment through poor diet, meaning foods that have been oxidized by processing or cooking, a build up of hydrogen ions results. The consequence of this is an acidic state and a low pH. To be healthy again, you need only change the environment of your body by adding alkaline substances that have an abundance of electrons such as *Ionized Water* and raw fruits and vegetables.

Ionized Water **reflects the characteristics of raw foods in several ways:**

- *Ionized Water* has an abundance of electrons as do raw foods.
- *Ionized Water* has a negative charge, or ORP, as do raw foods.
- *Ionized Water* possesses negative ions as do raw foods.
- *Ionized Water* is alkaline as are raw foods.

- *Ionized Water* is hydrating as are raw foods.
- *Ionized Water* is detoxifying as are raw foods.
- *Ionized Water* provides the body with ionic (organic) minerals as do raw foods.

Cooked foods have a deficit of electrons, a positive ORP and an abundance of positive ions. They are dry and dehydrating. They also acidify and add toxins to the body. All these qualities lead to disease and therefore are the exact opposite of what we should put in the body.

Ionized Water mimics many of the same attributes in nature that bring us health. And only nature can bring us true health. Vitamin and mineral supplements attempt to mimic nature although with dismal results. Medicine attempts to overcome the natural mechanisms of the body, thereby attempting to control, alter or outsmart it. Pharmaceuticals only mask symptoms of disease. No drug has yet been invented that cures the body of any disease, nor will one ever be. Pharmaceuticals never lead to health, but only allows people to hobble along a little further as they become sicker until the quality of their lives diminish to a point where the option of death is more appealing than continuing to live with such agony. True health is only found in nature. *Ionized Water* mimics and magnifies nature better than any substance known.

***Ionized Water* is for Everyone**

"Weed: a plant whose virtues have not yet been discovered."

~ Ralph Waldo Emerson

Ionized Water benefits everyone who consumes it. To get all the benefits of *Ionized Water*, one should drink it immediately, fresh from the tap. There are only a few retail locations in North America that sell *Ionized Water* by the gallon. While this may be a good way to introduce yourself to *Ionized Water*, ultimately you will want to purchase a water ionizer so you'll have fresh *Ionized Water* at your disposal. Having *Ionized Water* on tap is one of the best investments in your health you will ever make.

As *Homo sapiens*, we are more genetically identical than any other species on Earth. The variations in our individual genetic makeup are minuscule. These differences are less than 1/10,000 (10^{-4}) of a percent between individuals and are what give us different hair, skin and eye color, intelligence, blood type as well as facial and body shape. Therefore, we should all consume *Ionized Water* because it benefits every human in essentially the same way.

What to Expect from Drinking *Ionized Water*: A Synopsis

"Here are your waters and your watering place.
Drink and be whole again beyond confusion."

~ Robert Frost

The first thing that *Ionized Water* does to the body when a person begins drinking it is flush out the digestive tract, which is the best place for detoxification to start. This cleansing alone will significantly improve your potential for better health. Without a clean digestive tract, removing toxins from bodily tissue is much more difficult.

Collectively, digestive diseases are more common than any other kind of disease other than heart disease and cancer.[88] Since all chronic disease arises from our diet, it should not be surprising that the diseases that most commonly afflict us would be found in the digestive tract.

The body cannot absorb nutrients if the digestive tract is not clean. Enzymes chemically attack food to break it down throughout the digestive process until it is small enough to pass through the lining of the small intestines and into the bloodstream. Once there, the nutrients are carried away to the liver and other body parts to be processed, stored and distributed. None of this happens if the digestive tract is not clean of debris.

Living on a cooked-food diet over many years creates a layer of mucous throughout the digestive tract referred to as *mucoid plaque*, which is the hardening of mucous and other foreign materials created by living on a long-term cooked-food diet. This plaque coats the interior of the small intestines where nutrients are absorbed into the body. As it becomes thicker over the years we develop nutritional deficiencies even though we eat three meals a day that we believe are healthy. Our intestines become museums of partially digested, putrefying foods, drugs and other toxins that have never found their way out of the body. The intestines can swell to many times its size with only a pencil-size opening for feces to pass through. Even though we may be overeating, ironically we are starving ourselves under these conditions because the body is unable to absorb nutrients. Eating should never be confused with health. Many parents believe that as long as their child is eating they are healthy, and this certainly is *not* the case. All that matters when it comes to health is what foods we put in our body, not the frequency or quantity.

Ionized Water is a terrific cleanser of the digestive tract because of its small water molecule clusters. It penetrates the *mucoid plaque* that is attached to the intestinal lining helping to remove it. Once the surface of the lining is clean, the water then penetrates the lining itself to further cleanse and hydrate it so nutrients can pass into the blood with greater ease. When this happens, assimilation of nutrients to the body returns and the early stages of cellular rejuvenation can begin.

People comment how often they have to go to the bathroom when they first start drinking *Ionized Water*. This is because *Ionized Water* cleanses the digestive tract quite thoroughly. Anabolic, alkaline, small-cluster *Ionized Water* also travels through the body more quickly than conventional water. This problem will subside once your digestive tract is cleansed and your body becomes accustomed to drinking *Ionized Water*.

This Hurts: I Want to Stop

Detoxification

It's natural to feel pain that can come in many forms as poisons are excreted from the body. *Ionized Water* is the best water on Earth to consume, but it can bring great pain to those who are quite toxic. If you are in good physical condition, you won't have any problems adjusting to *Ionized Water*, but if you have a poor diet and don't exercise, you need to beware of *Ionized Water's* powerful detoxification properties. Water ionizers have various levels of strength so you can begin drinking it at a weak level (pH 7.6 – 8.0) and then increase its strength as you become acclimated to it (pH 9.5 – 9.9).

Begin by consuming one glass of *Ionized Water* on level one, the weakest level. Wait a couple hours and see if you begin to have any detoxification symptoms such as headaches, rashes, diarrhea and/or fatigue. When we detoxify, we are healing. If it becomes too painful, lower your consumption of *Ionized Water*, give it time and then begin drinking it at a stronger level. If you have any of these symptoms after drinking one glass of *Ionized Water* on level one, you are extremely toxic and need to introduce *Ionized Water* to your life very slowly. Your next glass of *Ionized Water* should be mixed with 50% filtered water only. Drink 2 – 3 glasses of diluted *Ionized Water* for one week then return to drinking straight *Ionized Water* on level one and see if any of the detoxification symptoms return. If they don't, slowly begin increasing the strength of the water daily until you have reached the highest pH level that should be consumed, which is 9.9. Of the *Ionized Water* that I consume, I drink as much of it as I can at pH 9.9 straight out of the tap.

Everyone's degree of toxicity varies, thus the effects of *Ionized Water* on your body will be slightly different than everyone else. It will depend on your historical and current diet, how much water you consume, how much you exercise, your mental attitude, how much medicine you take now and in the past, how stressful your life is and how toxic you are. If you continue to add toxins to your body through poor diet and/or a polluted environment this will prolong your cleansing.

If you don't have any detoxification symptoms on the lower ionization levels (pH 7.5 – 8.0), immediately jump to the higher levels and drink as much *Ionized Water* as you can each and everyday. *Never drink water around mealtime.* Other than that, water should rule your day.

How to Make Ionized Water Work for You

"The best way to prepare and balance drinking water so that it supports internal alchemy, stores and transmits energy, and promotes health is to first run it through a water ionizer to ionize the minerals, filter out toxins, remove acids, and alkalize the water."

~ Daniel Reid
A Complete Guide To Chi-Gung

Drinking *Ionized Water* needs to become part of your lifestyle if you expect to benefit from any its powerful attributes. *Ionized Water* can reverse the aging process to a degree, but only if you make it a regular part of your daily routine. Diet is the other change that must occur if you truly wish to reverse the biological aging processes so you are *internally* 20 years old at any age.

Hydrating. Nothing is more important for our health than a properly hydrated body. More disease comes from the failure to drink enough water than from any other source. When you start drinking *Ionized Water*, you will have more energy because your body is becoming better hydrated. You will also begin to think more clearly as the days and weeks pass. Water is the first lesson of health.

Alkalizing. Drinking *Ionized Water* reduces the body's acid level as it balances its overall pH. However, this will most likely take some time. Acid waste does not accumulate throughout the body overnight and it will not be flushed out overnight. Years of accumulated acid waste in the joints, around the organs, in the brain and throughout the body takes months and even years to completely expunge. As to how quickly you will rid the body of acid waste will also depend on your diet. The more acidic fried, cooked and processed foods we consume, the longer the alkalization process will take.

Oxygenating. Consistently drinking *Ionized Water* increases blood oxygen levels and therefore our energy level as well. Oxygen sharpens alertness because the brain is functioning closer to capacity when it is highly oxygenated. Our need for oxygen in an environment that is often oxygen-depleted due to pollution becomes of greater importance. *Ionized Water* is one of the best ways to oxygenate the body.

The Return of Thirst

"To underestimate one's thirst, to pass a given landmark to the right or left, to find a dry spring where one looked for running water - there is no help for any of these things."

~ Mary Austin

After you've been drinking *Ionized Water* for a while you will become much more sensitive to whether or not you are properly dehydrated. Your body's cries for water will become evermore intrusive on your life. Its many voices will scream louder until you satisfy its numerous demands. For example, I become aware that my body needs water after only an hour or so of not drinking any.

Simply by keeping my body constantly hydrated, I am accomplishing many things that are necessary to achieving Great Health. Staying hydrated helps stave off disease better than anything else we can possibly do for ourselves. When we are hydrated our blood is never too thin or too thick because of the lack of water or salt. We will have less of a tendency to retain salt or have high blood pressure when we drink enough water and include raw foods in our diet. Our organs, especially the kidneys, function optimally because they are not starved for water. Like any muscle, regular use of the kidneys only strengthens them. A common misconception is that drinking too much water can overtax the kidneys. If you have weak kidneys from living on a cooked-food diet, start by drinking weak *Ionized Water* (8.0 pH) then increase its strength slightly each day. When we provide the body with real nutrients from *raw foods* and drink enough water, weak kidneys can become powerful, flexing muscles once again, the way they were meant to be. This is true of any muscle or organ in the body.

In the first weeks of drinking *Ionized Water*, some people will not be able to quench their thirst. This is because your body is crying out for something it has rarely, if ever, experienced before: true hydration of its cells. The thirst mechanism in your body has been switched on for the first time in years, perhaps the first time in your life. We develop many ways to turn off our thirst or ignore it completely with foods or liquids that couldn't possibly hydrate us.

Proper hydration of the body is the beginning of Great Health.

What *Ionized Water* Has Done For Me: A Personal Testimonial

"I'm not obsessed. I'm just extremely preoccupied."

~ Anonymous.

I absolutely adore *Ionized Water*. I have since I consumed my first glass of it over 10 years ago. And because I am such a strong proponent of *Ionized Water,* I'm often asked what it has done for me personally since I began drinking it in 1996. Simply put, the effects on my body have been dramatic and profound, even though these positive changes occurred quite subtly. That is what everyone can expect from drinking *Ionized Water*: subtle yet profound changes in their health, all of them positive.

My life has been profoundly changed in positive ways since I began drinking *Ionized Water* and its gift to me has been Great Health.

I was already drinking a gallon of conventional water everyday when I began consuming *Ionized Water*. I also was a cooked-food vegetarian so I did not suffer the detoxification effects of *Ionized Water*, which are extremely powerful. Only those who are quite toxic will notice these strong effects. That's not to say I didn't detoxify because I did and it was a deep detoxification. I was a 95% cooked foodist at the time so I had a lot of cleansing to do. I did not exhibit the more brutal symptoms of detoxification that some do. However, we should all be cautious when we first begin drinking *Ionized Water*. Many people who believe they are healthy, including vegetarians and fitness gurus, suffer from the intense detoxification properties of

Ionized Water, which are like none other known. We must respect the power of tiny water molecule clusters because of their penetrating and cleansing properties.

I did get a few rashes in the early days as toxins came out of my body, but they were minor and I effectively treated them with *Acid Ionized Water,* which is excellent for treating rashes. I started by drinking *Ionized Water* at the highest level (pH 9.5 – 9.9) and consumed at least a gallon of it everyday. Now I drink 1.5 – 2.0 gallons of *Ionized Water* each day and I have never felt better in my life. You'll never have true energy and vitality without properly hydrating your body throughout each day. The subject of water should be the first chapter of every medical text book.

Since I was not ill when I first started consuming *Ionized Water*, I cannot claim that *Ionized Water* cured me of anything, although it has helped many others with a plethora of health conditions. Many people have written to me over the years testifying to the innumerable ways that *Ionized Water* has improved their health (Appendix 7: Testimonials). I have not kept them all, but if I were to have included them in this book, they would have easily doubled its size.

My body pH is balanced around 7.0 where it ought to be. My muscles are more relaxed because much of the lactic acid waste has been washed from my muscle tissue. I am almost 50 and do not ache in any of my joints. I am undoubtedly more limber, oxygenated, detoxified and better hydrated than any other time in my life.

My body's cells are bathed in the negative ORP of *Ionized Water*, which helps extinguish the oxidation of my body. Thus, I am retarding, if not reversing, the aging process at a cellular level.

I was in my late thirties when I started drinking *Ionized Water* and the many years of accumulated acid waste deposited throughout the human body takes months, even years to remove. It will take a lot of water and lot of time to fully accomplish the goal of complete pH balance and detoxification throughout my body.

But I could not have so easily gotten as far as I have without the daily consumption of *Ionized Water*.

Colds and flues have completely disappeared from my life since I began drinking *Ionized Water*. If I do feel a cold or sore throat coming on, I find that drinking a few extra glasses of *Ionized Water* quickly puts it at bay by hydrating and alkalizing my body so it can better fight the invasion.

Ionized Water is best used as a preventative health measure, although its healing properties are to be greatly respected as well. Water alone is the body's greatest ally for healing, better than anything a physician could possibly do for it. *Ionized Water* has many times the healing properties of conventional water. However, it is the body and only the body that is capable of healing itself of any disease. All we can do is provide it with the right ammunition so it can provide us with health. *Ionized Water* should definitely be in the arsenal we provide to the body for its defenses.

There are **Seven Components** of Great Health:

1. *Ionized Water*
2. Spirulina and Chlorella
3. Consumption of Probiotics (friendly bacteria)
4. Raw Foods
5. Angstrom Minerals
6. Vigorous Daily Exercise
7. Maintaining a Positive Mental Attitude (Spiritual Life)

Ionized Water is by far the most beneficial and important component. Making it an integral part of my life was the single best health decision I ever made.

Exercise and Fitness Enhancement

"Accept the challenges so that you can feel the exhilaration."

~ George Patton

Since I began drinking *Ionized Water,* it has provided me with an overall increase in stamina and recovery time while exercising. It is especially noticeable during and after strenuous cardiovascular activities such as running. My stamina has increased 25 – 30% and my recovery time has decreased a proportional amount. In the past, pushing hard on a five or six mile run would leave me with sore, tight muscles and maybe even aching knees the next day. This never happens now because years of accumulated acid waste has been neutralized and washed from my joints and muscle tissue by the alkalinity of *Ionized Water*.

I can report absolutely no arthritic conditions in my body. I am also far more limber than I was before I began drinking *Ionized Water*, thus it is much easier for me to stretch. And I do not need to stretch nearly as long to loosen up before exercising. The small water molecule clusters of *Ionized Water* absorb into the body much quicker and with greater efficiency. Muscle tissue that is properly hydrated will flex, work more efficiently and closer to its capacity when it is properly hydrated. In essence, we are stronger when we are hydrated because our muscles are able to work harder without straining. *Ionized Water's* superior hydration properties allow the body to perform to its capacity.

I have practiced Tae Kwon Do for years and all my injuries from that are all completely healed. I tripped while running and broke my ankle several years ago. The hair-line fracture healed and I was back running in ten days! While I must attribute part of the rapid healing to the powerful raw foods that I regularly consume, it would not have been possible if my body wasn't properly hydrated as well. Sufficient bodily hydration facilitates and encourages every aspect of healing and bodily repair whether it is from injury or damage from a disease such as cancer or osteoporosis.

In a competitive environment this puts an *Ionized Water* drinker at a distinct advantage for several reasons. The body operates most efficiently when it is alkalized, properly hydrated and its tissue is free of toxins. And while drinking conventional water just before a workout is ill-advised because of the potential for painful air pockets and the unpleasant feeling of water slogging around in your stomach, drinking *Ionized Water* is great before a workout. This is because the smaller water molecule clusters, or *microclusters*, absorb into the body's tissue with much greater efficiency. This doesn't allow the air pockets that cause side-aches to form inside the stomach and small intestines. Consuming strong, fresh *Ionized Water* (pH 9.5 – 9.9) just before exercising also provides the body with oxygen, which translates into extra energy.

Bodybuilders commonly do two dangerous and unhealthy things to themselves: acidify and dehydrate their bodies. They acidify it through their diet and then deliberately dehydrate it before a competition to make themselves look better. *There is nothing worse for the body than acidifying and dehydrating it.* Bodybuilders are guiltier of dehydration than any other group of athletes. This is one of the reasons why bodybuilders suffer from kidney, liver, heart disease, as well as a host of other health problems at a relatively young age. Use of steroids can also lead to a host of health problems because steroids use is completely unnatural. And like all artificial substances, steroids cause chemical confusion once inside the body. None of the steroid-induced processes that occur in the body are natural nor are the diseases that result from their use.

A hard-working bodybuilder appears to be a picture of perfect health as his or her muscles ripple and flex. It is the human body sculpted into a work of art, but often that body is deteriorating within, rife with disease because of dehydration, acidification and the poisons that are put into it. It does not have to be this way and *Ionized Water* can help change that by making weightlifters and bodybuilders healthier by alkalizing and hydrating them.

The Health of the Brain: Command Center of the Body

"The brain is nature's supreme achievement."

~ Friedrich Nietzsche

The most important use of *Ionized Water* is to benefit our brain. No other species on Earth is sentient and we owe this blessing to our brain, thus we should take great care of it. Because my brain tissue and cranial fluids are always sufficiently hydrated, I think much clearer since I began drinking *Ionized Water*. The brain is often deprived of oxygen, which results in clouded thinking. Thus, the clarity of my thought processes has definitely improved, both my memory and analytical functions. The body will do anything the brain tells it to. A brain that is adulterated by the toxins of a cooked-food diet or environmental poisons will send out the wrong neurochemical messages, causing the body to malfunction. While we need to alkalize, hydrate and detoxify the body to achieve Great Health, it is even more important that we do the same for the brain because it is our command center.

Brain material is like a sponge soaking up anything that passes through it. The brain is approximately 60% fat. Toxins and other unwanted substances such as *trans* fats, a.k.a. partially hydrogenated oils, cling to the gossamer material that composes the brain and over time diminishes its capacity to function properly. A dehydrated brain is even more susceptible to toxicity. Brain chemicals such as serotonin, dopamine and endorphins are either over-produced or under-produced, causing misfires in the brain's synapses. The result is a system-wide collapse of the brain's delicate balance, which can lead to a host of neurological and/or psychiatric problems ranging from schizophrenia, epileptic seizures, Parkinson's disease, dementia and even Alzheimer's disease. An elderly person with a lifetime of toxins accumulated in the brain is quite susceptible to such diseases, which is why more than half the people over age 85 fall victim to dementia-related, senility diseases. An alkalized, hydrated brain produces neurotransmitters in the correct amount and at the right time so that the brain functions properly. One of the best ways to flush these toxins from the brain is with *Ionized Water* because of its incredible hydration, detoxification and alkalization properties.

Not "Normal"

"There is something amiss with your reasoning."

~ Plato

The measurement of the concentration of particles in the urine is called *specific gravity*. The higher the number, the more particles are in the urine. The *specific gravity* of my urine generally falls out of the "normal range" (1.002 – 1.028) when measured because of the large amount of water I drink each day. However, the "normal range" is a misnomer in this case. Rather, it should be called the "typical range" because people typically do not drink enough water therefore their urine has more particles in it. When I notice my urine color darkening I know it is time to drink more water. There are other "normal ranges" that I probably would not fall into because the medical establishment is not use to seeing truly healthy people and over time has adjusted its parameters to accommodate a population that does not drink enough water nor consume the right foods. When you begin drinking a lot of *Ionized Water* on a regular basis, you too will begin to fall out of the "normal range" and into the healthy range.

Allergies and *Ionized Water*

"The American public is not aware that there might be potential allergenic and toxic reactions. With regular food, at least people know which foods they have an allergy to."

~ Jeremy Rifkin

As they do with disease, most people assume the reason they have allergies is because of their genes. Thus, they believe that it is inevitable that they will suffer from them. *Ionized Water* can be of significant help to allergy sufferers. Drinking *Ionized Water* helps keep the body alkaline, well hydrated and cleansed with penetrating, detoxifying water that flushes accumulated pollen, mold and chemical substances from the body. Consuming *Ionized Water* also helps remove mucous from the body. Mucous is formed from the chronic consumption of cooked foods, especially rice, wheat and processed foods that contain high amounts of gluten. Dairy products, especially those high in fat, produce the greatest amount of mucous in the body. Dairy products are the unhealthiest of all the food groups that are still considered to be healthy by the general population. It is amazing that in the 21st century most nutritionists and dietitians still consider dairy products to be essential for a healthy diet even though one-third of the population is lactose-intolerant, meaning that they cannot digest dairy products of any kind. Although lactose-intolerance is considered by the medical establishment to be a curse, it is a great blessing in regards to health. Another myth about dairy products is that they provide calcium to the body. Denmark, Norway and Holland consume the *most* dairy products in the world and have the *most* cases of osteoporosis, bone disease, heart disease and breast cancer.[89] Countries such as Gambia where almost no dairy products are consumed have the highest bone density in the world.[90]

Mucous locks allergens such as pollen, mold and synthetic chemicals into the tissue where they become deeply embedded and the discomfort they cause increases and intensifies. The result of this is what we refer to as "allergic reactions" such as sneezing, itchiness, coughing, redness, hives and swelling. All allergies arise in the body from it being placed in the wrong environment through poor diet and chronic dehydration of the body and its cells.

I have suffered from seasonal allergies since I was a child. In fact, I nearly died at the age of 10 from a severe reaction to ragweed pollen. I was not born with that allergy, but rather developed it through years of living on the wrong diet. My allergies persisted into adulthood and seemed to worsen as I got older. However, once I gave up dairy products, they subsided substantially. They still plague me to this day although to a small degree compared to when I lived on a cooked-food diet because my body was acidic and I didn't drink enough water.

People who suffer from sneezing, itchy eyes, scratchy throat, etc. during allergy season should use strong *Alkaline Ionized Water* (pH 10.0) to flush their eyes, nose, mouth, throat, ears and face in order to alleviate these unpleasant reactions. *Alkaline Ionized Water* is effective in relieving such reactions because these types of seasonal allergies are caused by pollen, which is the seed of a plant, and *Alkaline Ionized Water* retards plant growth because most plants thrive in a slightly acid, not alkaline environment.

Since *Acid Ionized Water* enhances plant growth, the result of applying *Acid Ionized Water* to the eyes and other areas of the body that are itchy or irritated due to pollen is to actually stimulate or waken the pollen, thus worsening the condition. The hydrogen ions and low pH of *Acid Ionized Water* is a conducive environment for most pollen. Flushing the eyes and other affected areas with *Alkaline Ionized Water* has the opposite effect that *Acid Ionized Water* has of actually allaying the allergic condition by suppressing the stimulation of pollen. It does this, in part, by alkalizing the environment.

Medications and *Ionized Water*

"Health consists of having the same diseases as one's neighbors."

~ Quentin Crisp

My body is always in a position of health where it does not require medications and *Ionized Water* is a critical component of my health protocol that helps me achieve that. Drinking *Ionized Water* while you are taking prescribed medications will not interfere with the function of most medications. However, medications should **not** be consumed with *Ionized Water* because its ionic charge may cause the medication to enter the bloodstream quickly and some medications are designed to be time-released. Thus, caution should be exercised in this regard.

Instant *Ionized Water*: The Magnesium Stick

Products are available that add a magnesium catalyst to conventional water that causes a mild ionization effect, resulting in a pH of 8.5 and an ORP of -50 mV. Placing the magnesium catalyst in a small bottle of conventional water and shaking it for 10 seconds creates a mild form of *Ionized Water*. It takes about three minutes for the ionization to fully take place, although it will begin immediately. The magnesium catalyst produces a charge when it is shaken that results in the production of hydroxyl ions and a higher pH.

Although the *Alkalized Ionized Water* produced this way is not nearly as powerful or healthy as that produced by an electric water ionizer, it is an acceptable substitute while traveling. One magnesium stick 3 inches long creates 300 ml (10.1 ounces) of *Alkalized Ionized Water*. It can be reused 800 – 1200 times, depending on the source water. Magnesium ionization sticks are a good way to introduce people to *Ionized Water*.

Bottling Ionized Water

A few companies have attempted to bottle and successfully market *Alkaline Ionized Water* through health food stores. However, bottling *Ionized Water* is a mistake because its antioxidant properties, the centerpiece of *Ionized Water*, are long gone by the time it reaches the shelf, as is most of its alkalinity. Bottled *Ionized Water* tastes a little smoother than conventional water because some of the reduced molecule clusters will linger for months after it is bottled. However, the overall strength of bottled *Ionized Water* is only a fraction of the power of *Ionized Water* fresh out of the tap. People who drink bottled *Ionized Water* often get a bad first impression of what it's actually like because it has reverted back to conventional water for the most part by the time they drink it. Practically speaking, bottling *Ionized Water* does nothing except give it a bad reputation.

How Long *Ionized Water* Lasts

Alkaline Ionized Water is not exceedingly stable. *Acid Ionized Water* is slightly more stable than *Alkaline Ionized Water*. *Ionized Water* gradually loses its charge depending on how it is stored. The best way to store *Ionized Water* is in a glass container placed in a cool dark place. The next best container to use is polycarbonate plastic, because it has the lowest potential to leach plastic into the water. Next best is PET (*polyethylene terephthalate*), followed by polypropylene. Finally, there is HDPE, which is the white plastic that milk is bottled in and has the greatest tendency to leach plastic into the water, therefore I recommend avoid HDPE if possible. In the end, filling up your car with gasoline exposes us to benzene, trihalomethanes and a host of other pollutants more toxic than plastic.

Keep *Ionized Water* out of the sunlight because the ultra violet rays will react with the ions in the water and weaken it. If stored in a container, *Ionized Water* will lose its strength more quickly once the container is opened and exposed to the air. Anything reacting with it will weaken its ionization and cause it to revert back to conventional water, which is why I preach that fresh is best, although I don't go anywhere without a bottle of *Ionized Water* because even days after it is produced, *Ionized Water* is still vastly superior to conventional water.

- **Hydroxyl Ions**: 10 – 20 minutes *(A small number will remain up to 24 hours)*
- **Negative ORP** (mV charge): 18 – 24 hours
- **Alkalinity** (high pH): 3 – 8 days
- **Smaller molecule clusters**: 4 – 18 months

Adding a magnesium alkaline stick will make the ionization last approximately 10 – 30% longer, depending on conditions and the quantity of water used with it.

Acid Ionized Water*:* Up to 150 days if stored in a cool, dark place, unopened. Exposing Acid *Ionized Water* to the air:

- **Hydrogen Ions**: 30 – 90 minutes *(A small number will remain up to 24 hours)*
- **High Positive ORP** (mV charge): 48 – 96 hours
- **Acidity** (low pH): 7 – 14 days
- **Smaller molecule clusters**: 8 – 24 months

How Much to Drink

"All or nothing."

~ Henrik Ibsen

The pH level of *Ionized* Water can be adjusted between 7.5 and 11+ with a water ionizer, depending on source water conditions. At pH of 10, *Ionized Water* begins to taste salty, so there is a limit to how strong *Ionized Water* should be consumed. It also becomes too caustic to consume at a pH higher than 10. A pH 9.5 – 9.9 is strong enough to satisfy the body's needs for oxygen, alkalinity, a negative ORP, electrons and proper hydration.

I start each day with a 16 oz. glass of *Ionized Water*, fresh out of the tap at a pH 9.5 – 9.9. Considering that we spend 6 – 8 hours each night sleeping, the first thing we should do upon waking in the morning is drink a big glass of water. After my first glass of *Ionized Water*, I draw another and drink it over the next 30 – 40 minutes. Therefore, each day I consume one liter of water before I put anything else in my body.

I never drink water 30 minutes before, during or 30 – 60 minutes after I eat anything. Drinking water with a meal dilutes the digestive process by weakening the stomach's hydrochloric acid. It also washes out digestive enzymes and the friendly bacteria (probiotics), not allowing them to attach to the food to break it down so its nutrients can be absorbed and utilized by the body. *Ionized Water* is particularly bad for digestion because it is so alkaline and the stomach needs an acid environment in order to properly digest the food that we put into it.

However, the notion that the stomach is in a constant state of extreme acidity is wrong. A healthy stomach produces hydrochloric acid (HCl) only as a result of chewing, not any other time unless an imbalance exists in the body that causes it to produce HCl when it shouldn't. All imbalances such as these are a result of a diet comprised primarily of cooked foods, especially fried and junk foods. Since these foods are acidic, the stomach remains acidic for long periods of time. Since raw foods are alkaline, they help bring the pH of the stomach upward toward neutral.

Depending on age and weight, I recommend the average person drink the following quantities of *Ionized Water* each day:

- **Infants**: One liter.
- **Young children:** 2 – 4 liters.
- **Teens**: 3 – 6 liters.
- **Adult**: 90 – 150 lb: 4 liters (1 gallon+)

- **Adult**: Over 150 lb: 6 liters (1.5 gallons+)

Review: The Miraculous Benefits of *Alkaline Ionized Water*

- Gives you energy.
- Provides the body with lots of oxygen.
- Provides the body with a negative charge (ORP) that counteracts the oxidation of the body.
- The negative ORP rejuvenates the body at a cellular level, reversing the biological age of the body.
- Removes accumulated acid waste and toxins from the body.
- Promotes overall health and healing by bringing the body into pH balance.
- Hydrates the body up to six times more effectively than conventional water.
- Minerals that are ionized are more easily assimilated by the body because of their ionic charge.
- Powders, such as flour, mix more thoroughly and smoothly.
- Promotes regularity and digestive health by internally cleansing the body.
- Helps relieve seasonal allergies.
- Extremely detoxifying with its negative ionic charge and small hexagonal water molecule cluster size.

The Importance of Fasting

"The freedom and ease you experience during fasting enables you to discover new undreamed depths to the meaning of life."

~ Herbert Shelton

What naturally follows a discussion about water is a discussion about fasting. Fasting is absolutely the *best* thing you can do for your health at any given time. If you are sick, fasting is the quickest way back to health, a shortcut that allows the body to focus on one thing: not digesting food or taking in nutrients, but that of healing itself. Digestion requires a lot of energy from the body. *"Fasting is the great remedy. The physician within!"* [91] The best fast you can do is a water fast and the best water for fasting is *Ionized Water.*

Fasting gives the body a chance to rest, cleanse and repair itself. Every major religion in the world promotes fasting. Jesus, Buddha, Mohammed, the prophets and many other spiritual figures through the ages fasted regularly.

You can fast for any length of time over six hours, which would be considered a mini-fast. Fasting one day a month is a great habit to develop. I fast for 24 hours quite often by eating a meal in the evening and then not eating again until the same time the following day. The habit of eating one large meal a day, usually in the late afternoon or early evening, was commonly practiced by ancient warriors, including the Romans, especially before a battle. Fighting on an empty stomach was easier, in part because the body is not burdened with the energy demands of digestion.

The word breakfast means to *break a fast*. Instead of eating something for breakfast, I continue my fast for as long as possible by drinking only water in the morning and often into the early hours of the afternoon. Therefore, I fast for at least 14 hours each day.

Fasting cleanses the body as well as the brain. Clear-minded thinking occurs when the stomach is clear; clouded thinking occurs when the stomach is full.

There are many types of fasting and many ways to fast. For instance, one can fast with *spirulina* and *chlorella*, which is extremely cleansing and energizing. I fasted on *spirulina* and *chlorella* for 10 days and never had so much energy in my life. A raw juice fast will produce similar results, but is not as intense or cleansing. There are green-vegetable fasts and citrus-juice fasts, but these are not true fasts because we are still putting food into the body. The only *true fast* is a *water fast*.

What you fast on will determine how quickly you expel toxins from your body. For instance, a juice fast will remove 2 – 20 days of accumulated toxins in a day, whereas an *Ionized Water* fast will remove up to 100 days of accumulated toxins in a single day.

If you experience severe headaches, rashes, cramps, stomach or joint pain, dizziness or nausea, then you are detoxifying too quickly and you need to slow down. The rewards of fasting are many, but it can be a rigorous and painful experience for those who are quite toxic. If you don't feel good while fasting or the pain becomes too intense, then eat something and the pain will subside. You will know best when it is time to pull back on the reins.

Detoxification stops immediately when we eat cooked foods of any kind. Cooked foods contain toxins that further poison the body. By eating cooked foods the body has no choice but to stop ridding itself of substances that do not belong in it because more are being added to it.

If you don't feel that you can go it alone, there are clinics and fasting retreat centers that will help you get through a fast. Some people recommend that fasts be supervised.

However, if you decide to fast, try to get some support from those around you. Ask them not to encourage you to eat something even though you may be hungry. After fasting for a couple days, your hunger will disappear, although you will still miss the habit of simply putting something in your mouth. Without a doubt, fasting is the best way to jumpstart your health. The best recommendation I have for anyone who is sick, regardless of the disease, is to fast.

Chapter 5

Acid Ionized Water

The Other Half of Miraculous

"Acid (Ionized) Water is generally best for bathing as it acts as an antiseptic for the skin."

~ Dr. David Jubb

Having *Acid Ionized Water* on tap is worth the price of a water ionizer in and of itself because of its extraordinary uses and benefits.

When *Acid Ionized Water* is freshly produced, small bubbles are present in the water, but not as many as on the alkaline side, which tends to be much cloudier with hydroxyl ions. The bubbles on the acid side are hydrogen ions, which are free radicals, which is why *Acid Ionized Water* should never be consumed. *Acid Ionized Water* is also an oxidant because it has a high positive charge or ORP (+700 to 800 mV), which is detrimental to our health. We must avoid substances that encourage oxidation such as cooked foods and *Acid Ionized Water*.

If acid minerals such as phosphorous or chloride are not in great abundance in the source water, the pH of the *Acid Ionized Water* will not drop significantly. However, even without the presence of a low pH *Acid Ionized Water* is still effective although the lower the pH, the stronger and therefore more effective it is. A residential water ionizer can achieve a pH of 4.0 – 6.5, depending on the mineral content of the source water. As with *Alkaline Ionized Water*, the water temperature and flow rate through the water ionizer are components that will also determine its strength.

It is easier to demonstrate the effects of *Acid Ionized Water* than those of *Alkaline Ionized Water* because they tend to be more immediate.

Plant Growth: Our Counterpart in Nature

We have a symbiotic relationship with trees and plants of every kind. We breathe in the oxygen that they exhale while they breathe in the carbon dioxide that we exhale. A plant's growth and health are significantly enhanced with the regular use of *Acid Ionized Water* and these benefits are unmistakable. After witnessing the effects of *Ionized Water*, you know that it works and you know it without any doubt in your mind.

I gave a flower to someone once and it wilted within an hour because it did not have any water. She was quite disappointed so I suggested we put the flower in a glass of *Acid Ionized Water*. Within two hours, the wilted flower had returned to its previous condition of rigidity, vibrancy and health. When she saw it, she instantly became a believer in *Ionized Water*, for she had witnessed a minor miracle. Something that we thought was on its way to the grave had indeed been resurrected.

Acid Ionized Water makes plants grow extremely well. Its low pH properties are quite favorable to plants that usually thrive best if the soil is slightly acidic. The adjustability of the pH on a water ionizer is helpful in this regard. The smaller water molecule clusters of *Acid Ionized Water* transpire into the plant much more effectively than conventional water, which helps increase the turgor of the plant. *Acid Ionized Water* contains an abundance of hydrogen ions as do many fertilizers. Thus, watering a plant with *Acid Ionized Water* has an effect similar to that of fertilizing it.

The difference between feeding a plant *Acid Ionized Water* and conventional water is profound. When plants are fed *Acid Ionized Water*, it becomes apparent when it first started receiving it by the healthier growth from that point on the plant up as it grows taller. Overall, plants that are watered with *Acid Ionized Water* look much healthier. Their flowers will be brighter, their leaves shinier with richer coloring. *Acid Ionized Water* can also be used effectively on lawns and golf courses as a combination fertilizer and pesticide.

Retards Bacterial Growth

Acid Ionized Water kills bacteria on contact. How effective it is depends on the strength of the *Ionized Water*, meaning how high the ORP (mV) and how low the pH are. For instance, an ORP of 1100 mV and a pH of 2.5 is considered anti-microbial, meaning that it kills all bacteria on contact. By comparison, chlorine, which is also an oxidant, has a one-second bacterial kill rate at 700 mV. Mild *Acid Ionized Water* produced from tap water using a home water ionizer typically has a range of 700 – 950 mV, which is quite effective at killing bacteria and retarding its growth. As with *Alkaline Ionized Water*, the strength of *Acid Ionized Water* will depend on the mineral content, temperature and flow rate of the source water through the water ionizer.

To demonstrate *Acid Ionized Water's* anti-bacterial abilities, place two oranges side by side and spray one with strong *Acid Ionized Water* three or four times a day and not the other. The orange that is not sprayed with *Acid Ionized Water* will decay much faster than the one sprayed with the *Acid Ionized Water*, which will last several days longer. *Acid Ionized Water* kills the bacteria on the sprayed orange, thus allowing it to decay more slowly than the one that was not sprayed with it.

Skin, Hair and Scalp Conditioner

Acid Ionized Water has a wonderful conditioning effect on the skin and hair because they both are somewhat acidic. Acidic skin is the body's first line of defense against bacteria. Human skin typically has a pH of 5.4 in order to ward off bacteria and infection, and hair typically has a pH of 5.6, although these can vary slightly between individuals.

Applying *Acid Ionized Water* regularly to the skin works as an astringent to tighten it and help remove wrinkles, leaving no chemical residue as with other astringents. The only residue on the skin is water, which is harmless. *Acid Ionized Water* soothes and helps keep the skin clear of acne and other blemishes. Since acne is caused by bacteria (*propionibacterium acnes*), *Acid Ionized Water* helps destroy the cause of the acne; and because of its smaller molecule cluster size, it is able to penetrate into skin pores more effectively to kill the bacteria.

Skin and hair, even animal fur, respond positively to the conditioning effects of *Acid Ionized Water*. The more *Acid Ionized Water* that is applied on the skin and hair, the better they respond to it and there is no limit to the number of applications one can have each day.

Combined with drinking *Ionized Water*, which alkalizes the body internally, skin conditions of every kind such as *psoriasis* and *shingles* are dramatically improved. However, this can be short-lived if the root of the problem is not addressed by a fundamental change in diet. For instance, a condition such as *psoriasis* arises from a poor diet of junk foods and excessive animal protein. Chronic skin problems are a result of toxins making their way to the skin on their way out of the body. For instance, leaky gut syndrome is a disease where proteins are deposited through the stomach wall and directly into the bloodstream before they have been digested and processed so they can be used by the body. Undigested proteins in the bloodstream are essentially toxins that need to be removed from it. The result is skin rashes that surge and ebb, but are never resolved, which becomes a lifestyle problem that most people believe is something they simply must life with. People with chronic skin conditions go through life with ointments and creams that allay the symptoms of their disease without ever treating their cause. If the diet is not changed to include more alkaline foods, *Acid Ionized Water* will only temporarily relieve skin conditions such as *psoriasis*.

Rashes, cuts, scrapes, even serious wounds, as well as Athlete's Foot and other fungus are dramatically improved with the application of *Acid Ionized Water*. It takes the itch out of mosquito bites and alleviates the pain of stings and other insect bites. It provides relief from poison oak and poison ivy exposure. It will even take down the swelling of a swollen lip. It can used for battery cell water with excellent results because of the strong conductivity due to the high positive mV (ORP) nature of the water. A catalytic environment is perfect for quick reactions such as what is required from a car battery. Catalytic reactions burn hot and fast. Anabolic reactions burn cool and slow.

Acid Ionized Water acts as an astringent, helping to tighten skin and remove wrinkles, yet without any chemical residue. Dry skin is best treated by soaking it in *Acid Ionized Water* for 20 – 30 minutes. The deep cuts near the fingernail that people experience in extremely cold, dry weather will heal in a few days after soaking them in *Acid Ionized Water* for 20 – 30 minutes.

Scalp problems such as dandruff are improved with the consistent use of *Acid Ionized Water.* Eating a proper diet comprised of spirulina, chlorella, raw fruits and vegetables, as well as drinking *Alkaline Ionized Water*, can entirely eliminate many of these health problems. Health problems are an indication that key nutrients in the diet are missing and that the body is not properly hydrated.

Acid Ionized Water has been used successfully in treating diabetic skin ulcers, wounds that open up on the skin, particularly the extremities, due to poor circulation. These ulcers are difficult to heal because they do not easily cauterize or drain properly. Diabetics sometimes must have feet and even legs amputated because the ulcers become infected with gangrene and are beyond treatment. However, soaking these ulcers in *Acid Ionized Water* has a tremendous healing effect on them. It kills the bacteria around the wound and allows it to cauterize so it can then heal. Doctors insist that diabetic ulcers must be kept absolutely dry at all times, yet *Acid Ionized Water* is an effective treatment which defies that rule with tremendous results! Medical professionals are at a loss to explain the success of using water to treat diabetic ulcers.

Drinking *Alkaline Ionized Water* is a great benefit to diabetics because it helps bring the body into pH balance and provides it with lots of oxygen, which increases circulation. This encourages the transportation of nutrients around the body and further brings the body into overall balance, or *homeostasis*, which is key to fighting any disease. Once again, changing the environment of the body with negatively charged hydroxyl ions helps bring it into a position where it can defend and heal itself.

Applying Acid Ionized Water

Acid Ionized Water should be applied directly to the skin for it to work effectively. The stronger *Acid Ionized Water* is, the more effective it is. It can be dabbed, splashed or sprayed on and allowed to dry. For the best results, it should be used on the skin, fresh and as often as possible. The most effective way to use *Acid Ionized Water* is directly out of the tap when applying it to your skin or brushing your teeth with it. *Acid Ionized Water* kills the bacteria in the mouth, facilitates the removal of plaque from the teeth, and soothes and promotes gum health. The smaller water molecule clusters penetrate areas between the teeth and at the gum line where bacterial growth can cause cavities. I personally have not used toothpaste since I discovered *Ionized Water*. I find *Acid Ionized Water* to be just as effective as toothpaste for cleaning my teeth and encouraging gum health.

Adding a gallon or more of *Acid Ionized Water* to your bath water will give you softer skin. Or you can pour it over your head after a shower for a similar effect. The uses for *Acid Ionized Water* as a topical treatment are limited only to your imagination.

The Miraculous Benefits of Acid Ionized Water

- Kills or retards growth of bacteria on contact.
- Helps heal cuts, blisters, scrapes, rashes, burns, serious wounds.
- Helps heal diabetic skin ulcers.
- Provides relief from mosquito bites and bee stings.
- Provides relief from poison ivy and poison oak rashes.
- Cleans and conditions hair.
- Great for car, truck or boat acid batteries.
- Relieves chapped hands and dry, itchy skin.

- Effectively removes plaque from teeth. Use it instead of toothpaste.
- Wash vegetables, fruits, meat and fish with it to kill bacteria.
- Gargle with it to relieve sore throats and mouth sores.
- Acts as an astringent to tighten skin and remove wrinkles.
- Leaves no chemical residue when using it on the skin.
- Helps retard the spread of acne, eczema and other skin and scalp conditions.
- Excellent for treatment of fungus, such as Athlete's Foot.
- Excellent cleaning agent.
- Promotes healthy plant growth.
- Increases plant turgor.
- Significantly extends the life of cut flowers.

Super *Acid Ionized Water*

Super acid water ionizers produce very strong *Acid Ionized Water* with an ORP of 1100 mV and a pH of 2.5 . These stronger attributes are achieved by using much larger electrodes than are found in residential water ionizers. Some super acid water ionizers add pure salt (*sodium chloride*) to the water, which drives up the pH on the alkaline side and drives down the pH on the acid side. Sodium is used because it is an excellent conductor. Using sodium with a residential water ionizer will cause it to overheat because of the excessive conductivity it creates in the water. Sodium is an alkaline mineral, chloride is an acid mineral. By using the sodium to raise the pH of *Ionized Water* on the alkaline side and chloride on the acid side, the result is very strong forms of both *Acid Ionized Water* and *Alkaline Ionized Water*. Super *Alkaline Ionized Water*, with a pH of 13.0 – 13.7, is an excellent degreaser, however is far too caustic for consumption even though its ORP may become as low as -1000 mV.

Super *Acid Ionized Water,* or HOP (High Oxidation Potential) water, is used primarily for medicinal purposes, such as advanced diabetic ulcers, gangrene, serious wounds or burns. At a pH of 2.5, it is antimicrobial, meaning that it kills all bacteria on contact. It is far too strong for plants and should not be used to water them. Super *Acid Ionized Water* could be used as a hand disinfectant in hospitals where 80% of all people who are infected while in the hospital contract their disease by way of doctors and nurses not washing their hands between patient visits. Using it to wash your hands at a pH of 4.0 would kill essentially all the bacteria, thus contributing to a cleaner hospital environment. However, direct contact with HOP water below a pH of 3.0 can cause excessive drying of the skin.

HOP water could also possibly be used with livestock and especially on newborns, which would promote a cleaner working environment and help lower infant mortality rates and veterinary costs. Since most applications are intended for commercial purposes, they have limited practical uses for the average household.

Book Summary

"What's water but the generated soul?"

~ William Butler Yeats

Water is our best defense against disease of every kind. Sixty percent or more of all chronic disease would be significantly reduced if people would simply keep themselves properly hydrated.

To ionize means to gain or lose an electron. Essentially, the ionization process robs an electron from one molecule and donates, or transfers, it to another molecule. Both *Alkaline* and *Acid Ionized Water* have extraordinary properties and benefits; however, their respective uses could not be more different. We consume *Alkaline Ionized Water* and use the *Acid Ionized Water* on the outside of our bodies for acne, cuts, scrapes and rashes of all kinds. It kills bacteria on contact and encourages plant growth.

The centerpiece of *Alkaline Ionized Water* are its antioxidant properties. It is miraculous that normal tap water can be instantly transformed into a strong antioxidant. *Alkaline Ionized Water* has two antioxidant qualities; its negative charge and the presence of hydroxyl ions which are free radical scavengers. The body is starved for electrons and *Alkaline Ionized Water* contains an abundance of them, which nullify free radicals in the body.

Alkaline Ionized Water is an extremely effective antioxidant because it is a liquid that has a small grouping of water molecule clusters and thus is more easily absorbed into the body where it can be of immediate use. Drinking *Alkaline Ionized Water* gives you energy through better hydration and alkalization of the body and by providing the body with oxygen.

Because of the predominance of hydroxyl ions in *Alkaline Ionized Water*, the water becomes alkaline, meaning it has a high pH. The pH level can be adjusted with a water ionizer between 7.5 and 9.9, which is the highest pH that it should be consumed.

All disease thrives in an acid environment in the body and will not flourish or thrive in an alkaline environment. If we acidify our bodies through poor diet we become vulnerable to any disease that invades the body. The more acid our body is, the more susceptible we become to disease.

Ionized Water is sometimes referred to as micro-cluster water because of its small molecular grouping. Water molecules typically group in clusters of 10 or more. *Ionized Water* molecule clusters group together into six water molecules, thus they have been reduced in size, which is the most natural state for water to exist in. Ionization changes conventional water from an irregular, bulky shape to a hexagonal shape that saturates body tissue much more efficiently. These smaller six-sided clusters are extremely penetrating and hydrating. As it hydrates body tissue, *Ionized Water* pushes out all the things that don't belong in the body, which are commonly referred to as toxins. Therefore, *Ionized Water* is extremely detoxifying, which is why people who are quite toxic must start drinking *Ionized Water* slowly so they do not detoxify too quickly.

Ionized Water is best consumed straight out of the tap because it is most effective when it is fresh. *Ionized Water* should become a part of everyone's lifestyle if they wish to be healthy. It is the best substance we can possibly consume because there is nothing healthier for the body than water and there is no better water than *Ionized Water*.

Appendix 1

Comparison between Waters

BENEFITS	Ionized Water	Bottled Water	Tap Water	Well Water	Purified Water
Powerful Antioxidant	**Yes**	No	No	No	No
Negative ORP	**Yes**	No	No	No	No
Balances Body pH, Alkalizes the Body	**Yes**	No	No	No	No
Powerful Detoxifier	**Yes**	No	No	No	No
Superior Hydrator	**Yes**	No	No	No	No
Enhances Mineral Absorption	**Yes**	No	No	No	No
Increases Oxygen	**Yes**	No	No	No	No
Low Cost	**Yes**	No	Yes	Yes	Yes*

* After the initial purchase of a distiller or reverse osmosis unit.

Appendix 2

The Use and Function of Water Ionizers

"Tomorrow is often the busiest time of the year."

~ Spanish Proverb

If you don't have one, the best time to purchase a water ionizer is immediately. You will not need a plumber to install the water ionizer. Most water ionizers will attach to any faucet in your house and take approximately 10 – 15 minutes to install. They are about the size of a large toaster (11" across x 12" high x 6" deep on average), take up very little counter space and most can be hung on the wall. You do not need to give up the use of your sink when you install your water ionizer because they come with a faucet diverter that attaches to the faucet, which allows you to use your sink as normal then divert the water into the water ionizer as needed.

Some water ionizers have a valve built into them so that they can be directly plumbed into your water line. However, nearly all water ionizers are designed to be used on top of the counter so they can drain between uses since stagnant water can potentially cause a bacterial hazard.

Water ionizers are currently manufactured only in Asia. There are several water ionizer models available on the market and most of them offer comparable performance, although retail prices can vary widely. Water ionizers are analogous to toasters in that prices may vary for a toaster, although regardless of how much you pay, you still end up with toast. The same is true of water ionizers. No matter what you pay, you still end up with *Ionized Water*.

Purchasing a Water ionizer: What to Look For

The basic requirement of any water ionizer is for it to have an ability to produce water with a pH between 9.5 – 9.9.

Electrodes. The size and number of platinum-coated titanium electrodes that the water ionizer has will determine the strength of the *Ionized Water* that is produced. More electrodes produce stronger *Ionized Water*. Water ionizers typically have between 3 – 7 electrodes that vary in size, depending on the manufacturer. The membranes between the electrodes are composed of either pulp mesh or complex plastic polymers. Polymer membranes are more likely to warp if exposed to high temperatures. Pulp mesh is more resistant to high temperatures than plastic polymers.

Transformer. The size of the transformer will also determine how strong the *Ionized Water* will become. Small transformers apply less power and will overheat quicker than larger ones.

Reliability. Preferably, a water ionizer model ought to be on the market for at least a year before you should consider purchasing it. This precaution should be taken because new units sometimes have design flaws that need to be fixed and their first year on the market is when they will most likely be found. The reputation of the manufacturer is also of great importance when choosing a water ionizer. I prefer to purchase a water ionizer from a company that manufactures nothing other than water ionizers, not only as a sideline product. Reputation is everything in regards to water ionizer manufacturing.

Ultraviolet Light. Buying a water ionizer with an ultraviolet light (UV) to disinfect the water as it enters the unit is an unnecessary expense. UV light is not needed in most cases and becomes ineffective after only a few months of use because the tube that carries the water films over with mineral deposits, reducing the effectiveness of the UV light.

Warranty. Water ionizers typically come with anywhere from a 1 – 5 year warranty. Make sure the company you purchase your water ionizer from has repair centers in the country where you purchase it. If they claim to come with a lifetime warranty, I suggest reading the warranty carefully. Lifetime warranties do not often come with electronic products.

Where and how to Install a Water Ionizer

Water ionizers can be installed at any sink in the house, most commonly in the kitchen or bathroom. Typical installation of a water ionizer takes approximately 10 minutes and does not require a plumber. Depending on the unit, most water ionizers come with a faucet diverter and about 5 – 8 feet of tubing. To install the unit, remove the aerator on the sink faucet and replace it with the faucet diverter. Unless it is metric, the faucet diverter that comes with your unit is a standard size 15/16" (27 thread) male fitting, which fits most standard faucets. If the threads are inside the faucet (female threads) you will need a 15/16" male adapter (27 threads - 9D), which can be purchased at any hardware store. If the threads are slightly smaller than the 15/16" faucet diverter, you may have a 13/16" faucet in which case you will need to purchase a 15/16" x 13/16" adapter. Less common is a 55/64" thread size, which requires a 15/16" x 55/64" adapter.

The most difficult faucets to attach a water ionizer to are designer faucets. Often they are simply not suitable for use with a water ionizer because the fittings are unusual sizes. Since there are dozens of these types of faucets on the market, manufacturers are unable to supply the fittings that may be required to install a water ionizer to them. If you are unable to find the proper fittings for your designer faucet, you may want to consider hooking up the water ionizer to another sink in your house, such as the bathroom or laundry room where there is a standard faucet.

Once you have attached the faucet diverter to your faucet, you can continue to use the sink as you normally do, then divert the water into the water ionizer when you wish to use it by turning the handle on the faucet diverter. Cut the single length of tubing in half and attach one piece from the diverter into the inlet of the water ionizer. The other piece of tubing attaches to the *Acid Ionized Water* fitting on the bottom of the unit and drapes into the sink. *Alkaline Ionized Water* flows from a stainless steel flexible tube on top of the unit, except when the unit is in **cleaning mode** and *Acid Ionized Water* is flowing from the silver flexible tube and

Alkaline Ionized Water is flowing from the bottom tube. Both *Alkaline* and *Acid Ionized Water* are created simultaneously when water is run through a water ionizer, approximately 70% *Alkaline Ionized Water* and 30% *Acid Ionized Water*.

Acid Ionized Water can be collected for use at a later time, although the fresher it is, the stronger and more effective it is.

Typical Water Faucet Installation

Step 1: Screw in stainless steel flex hose to the top of the unit.

Step 2: Remove the aerator from the faucet and attach the faucet diverter to faucet.

Step 3: Cut the plastic hose that comes with your water ionizer in half.

Step 4: Attach one hose to the faucet diverter using a hose clamp at each end of the hose to prevent leaks. (If your water ionizer has compression fittings, this will not be necessary.) Attach the other end of the hose to the bottom of the water ionizer marked INLET.

Faucet Diverter

Step 5: Attach the other hose to the bottom of the water ionizer labeled OUTLET or Acid Water and drape the tube into the sink.

Step 6: Plug the unit into an electrical outlet.

Step 7: Turn on the faucet and begin running water through the unit. The water ionizer will turn on automatically. The water will briefly turn a bluish color as the carbon powder is washed from the filter. After one minute of flushing, the *Ionized Water* is ready to consume.

Using Five Gallon Bottled Water with a Water Ionizer

If you wish to use bottled spring water with your water ionizer, you will need to purchase an electric bottled water dispenser to pump it from the 5 gallon bottles. The electric pump will allow you to pump water through your water ionizer from up to 20 feet away. All that is required is a 1/4" hose clamp to make the leak-proof connection between the water ionizer and the tube from the 5 gallon bottled water dispenser.

Cleaning Your Water ionizer

After several months, the electrodes on a water ionizer coat with minerals such as calcium, which will reduce the strength of the *Ionized Water* since the source water is no longer coming into direct contact with the electrodes. Thus, this layer of calcium begins to inhibit ionization from occurring to some degree. The reason for the mineral build-up on the alkaline side of the water ionizer is because the high pH of the water causes the minerals in the water to precipitate out of the water and become displaced on the electrodes. Running the unit in cleaning mode reverses the electrode polarity and will help slow this mineral depositing, but it will not entirely stop it from occurring. If there is a disproportionate amount of water coming out the tube that drains into the sink (the acid tube) and very little coming out the silver tube (the alkaline tube), the unit may need cleaning because of mineral build-up on the electrodes and the alkaline water tube on top of the unit. This typically does not become a problem until a unit has been in use 12 – 24 months.

Cleaning your water ionizer using this prescribed method will allow the *Ionized Water* to become stronger because the water is now coming into direct contact with the electrodes instead of the calcium that has deposited on them.

Directions for cleaning your water ionizer

1. Unplug the water ionizer from the electrical outlet. Remove as much water from the water ionizer as possible by blowing through the silver tube, inlet tube and acid tube. Also remove the filter and shake as much water from it as you can then replace it. It is preferable to use an empty filter that has no carbon media in it while cleaning the unit. An empty filter can be purchased from a water ionizer dealer.

2. Prepare a container with 2 liters of white vinegar or citric acid for the pump. The most common pump used for this is a fish tank pump, which can be purchased at any pet supply store. Attach the INLET hose from the water ionizer to the pump. Depending on the pump you purchase, this may require some minor modifications so that the hose will fit into the pump properly. Submerge the pump into the white vinegar or citric acid. Plug the pump into the electrical outlet. A GFI safety electrical outlet should be used for this in order to protect against electrocution.
3. Place both the silver tube and acid water tube from the water ionizer into the white vinegar or citric acid container so the entire system is recirculating.
4. At first, there may be very little flow through the silver tube. However, as long as the slightest amount of white vinegar or citric acid is flowing through the silver tube, it will eventually dissolve the calcium that is blocking the water from flowing out of the tube.
5. Clean the water ionizer for 2-12 hours depending on how much calcium and other minerals have been deposited on the electrodes. Reattach the water ionizer to your water source and flush the unit with water for about 10 minutes to clean out the white vinegar or citric acid from the unit. Keep unit unplugged until vinegar or citric acid taste has disappeared from water. When that is accomplished, your water ionizer is ready to be used again.

   Note: You can also remove the silver tube from the unit and submerge it completely in the white vinegar or citric acid overnight to remove the mineral build-up inside it.

Appendix 3

Top Ten Reasons to Never Consume Carbonated Soft Drinks

"Many Chinese saw opium as a poison introduced by foreign enemies."

~ Robert Trout

1. **Soft drinks are a water thief.** They are a diuretic that steals more water than they provide to the body. Processing the high levels of sugar in soft drinks removes a considerable amount of water from the body. To replace the water stolen from the body by soft drinks, you need to drink 8 – 12 glasses of water for every one glass of soft drinks that you consume.
2. **Soft drinks never quench your thirst, certainly not your body's need for water.** Constantly denying your body adequate amounts of water leads to *chronic cellular dehydration*, a condition which weakens your overall health. This, in turn, can lead to a weakened immune system and a plethora of diseases.
3. **The elevated levels of phosphates in soft drinks leach vital minerals from the body.** Soft drinks are made with purified water that also leaches vital minerals from the body. A severe lack of minerals can lead to heart disease (lack of magnesium), osteoporosis (lack of calcium), psychiatric disorders (lack of phosphorous) and many other diseases. Most vitamins cannot perform their function in the body without the presence of minerals.
4. **Soft drinks can remove rust from a car bumper or other metal surfaces.** Imagine what it's doing to your digestive tract as well as the rest of your body.
5. **The high amounts of sugar in soft drinks causes the pancreas to produce an abundance of insulin, which leads to a "sugar crash."** Chronic elevation and depletion of sugar

and insulin levels in the body can lead to diabetes and other hormonal imbalances. This is particularly disruptive to growing children and can lead to lifelong health problems.

6. **Soft drinks severely interfere with digestion.** Caffeine and high amounts of sugar virtually shut down the digestive process. This means your body is essentially taking in NO nutrients from the food you consume.

7. **Diet soft drinks contain Aspartame, which has been linked to depression, insomnia, neurological disease and a plethora of other illnesses.** The FDA has received more than 10,000 consumer complaints about Aspartame. This accounts for 80% of all complaints concerning food additives.

8. **Soft drinks are extremely acidic, so much so that they can eat through the liner of an aluminum can and leach aluminum ions from the can if it sits on the shelf too long.** Autopsies of Alzheimer patients have revealed high levels of aluminum and other heavy metals lodged in their brains. Heavy metals in the brain can lead to numerous neurological and psychiatric diseases. The human body is healthiest at a pH of about 7.0. Soft drinks have a pH of approximately 2.5, which means you are putting something into your body that is approximately 50,000 times more acidic that your body ought to be. All diseases flourish in an acidic environment.

9. **Soft drinks contain large amounts of benzene, eight times that which is allowed in drinking water.** Benzene is a carcinogenic chemical that has been linked to leukemia.

10. **Soft drinks are one of the worst substances you can possibly put in your body.** Don't even think of taking a sip of a soft drink when you are sick with a cold, flu or something worse. It will only make it that much harder for your body to fight the illness. This is because of the huge amount of sugar, chemicals and acid liquids you are consuming. Disease loves it, but your body will not.

Appendix 4

Top Ten Reasons We Need To Drink Water

"Water is a finite resource that is essential in the advancement of agriculture, and is vital to human life."

~ Jim Costa

1. **Water is the substance of life.** Life cannot exist without water. We must constantly add fresh water to our bodies in order to keep our cells properly hydrated.
2. **The body is comprised of approximately 69% water.** It is the most important thing we consume. Water is the cornerstone of health. More than 50% of the world's population is chronically dehydrated.
3. **It is difficult for the body to get water from any other source than water itself.** Soft drinks and alcohol steal tremendous amounts of water from the body. Even other beverages such as coffee, milk and juice require water from the body to be properly digested.
4. **Water plays a vital role in nearly every bodily metabolic function**, therefore we must constantly stay hydrated.
5. **Water is essential for proper digestion, nutrient absorption, enzyme performance and chemical reactions.**
6. **Water is essential for proper blood circulation**. Drinking water is also essential for healthy weight reduction.
7. **Water helps remove toxins from the body**, in particular from the digestive tract.
8. **Water regulates the body's cooling system.** A dehydrated body will overheat much quicker than a sufficiently hydrated one.
9. **Consistent failure to drink enough water can lead to Chronic Cellular Dehydration.** This is a condition where the body's cells are never quite hydrated enough, leaving them in a weakened state, vulnerable to attack from disease. It weakens

the body's overall immune system and leads to chemical, nutritional and pH imbalances that can cause a host of diseases.

10. **Dehydration can occur at any time of the year**, not only during the summer months when it is hot. The dryness that occurs during winter can dehydrate the body even quicker than when it is hot. Victims of diseases such as cholera often die primarily due to dehydration, not from the disease itself. Dehydration is a leading cause of mid-day fatigue because it lowers metabolism. It also causes foggy thinking and short-term memory loss.

Appendix 5

Scientific Studies of the Effects of *Ionized Water*

Ther Apher Dial. 2009 Jun;13(3):220-4.

Ionized alkaline water: new strategy for management of metabolic acidosis in experimental animals. Abol-Enein H, Gheith OA, Barakat N, Nour E, Sharaf AE.
Department of Urology, El Mansoura Urology and Nephrology Center, Mansoura University, 72 Gomhoria Street, Mansoura, Egypt.

Metabolic acidosis can occur as a result of either the accumulation of endogenous acids or loss of bicarbonate from the gastrointestinal tract or the kidney, which represent common causes of metabolic acidosis. The appropriate treatment of acute metabolic acidosis has been very controversial. Ionized alkaline water was not evaluated in such groups of patients in spite of its safety and reported benefits. So, we aimed to assess its efficacy in the management of metabolic acidosis in animal models. Two models of metabolic acidosis were created in dogs and rats. The first model of renal failure was induced by ligation of both ureters; and the second model was induced by urinary diversion to gut (gastrointestinal bicarbonate loss model). Both models were subjected to ionized alkaline water (orally and by hemodialysis). Dogs with renal failure were assigned to two groups according to the type of dialysate utilized during hemodialysis sessions, the first was utilizing alkaline water and the second was utilizing conventional water. Another two groups of animals with urinary diversion were arranged to receive oral alkaline water and tap water. In renal failure animal models, acid-base parameters improved significantly after hemodialysis with ionized alkaline water compared with the conventional water treated with reverse osmosis (RO). Similar results were observed in urinary diversion models as there was significant improvement of both the partial pressure of carbon dioxide and serum bicarbonate ($P = 0.007$ and 0.001 respectively) after utilizing alkaline water orally. Alkaline ionized water can be considered as a major safe strategy in the management of metabolic acidosis secondary to renal failure or dialysis or urinary diversion. Human studies are indicated in the near future to confirm this issue in humans.

Drug Dev Ind Pharm. 2009 Feb;35(2):145-53.

Effect of several electrolyzed waters on the skin permeation of lidocaine, benzoic Acid, and isosorbide mononitrate. Kitamura T, Todo H, Sugibayashi K.
Faculty of Pharmaceutical Sciences, Josai University, Saitama, Japan.

The effects of several electrolyzed waters were evaluated on the permeation of model base, acid and non-ionized compounds, lidocaine (LC), benzoic acid (BA), and isosorbide mononitrate (ISMN), respectively, through excised hairless rat skin. Strong alkaline-electrolyzed reducing water (ERW) enhanced and suppressed the skin permeation of LC and BA, respectively, and it also increased the skin permeation of ISMN, a non-ionized compound. On the contrary, strong acidic electrolyzed oxidizing water (EOW) enhanced BA permeation, whereas suppressing LC permeation. Only a marginal effect was observed on the skin permeation of ISMN by EOW. These marked enhancing effects of ERW on the skin permeation of LC and ISMN were explained by pH partition hypothesis as well as a decrease in skin impedance. The present results strongly support that electrolyzed waters, ERW and EOW, can be used as a new vehicle in topical pharmaceuticals or cosmetics to modify the skin permeation of drugs without severe skin damage.

J Toxicol Sci. 2000 Dec;25(5):417-22.

Influences of alkaline ionized water on milk electrolyte concentrations in maternal rats.

Watanabe T, Kamata H, Fukuda Y, Murasugi E, Sato T, Uwatoko K, Pan IJ. Department of Veterinary Biochemistry, College of Bioresource Sciences, Nihon University, 1866 Kameino, Fujisawa, Kanagawa 252-8510, Japan.

We previously reported that body weight on day 14 after birth in male offspring of rats given alkaline ionized water (AKW) was significantly heavier than that in offspring of rats given tap water (TPW), but no significant difference was noted in milk yield and in suckled milk volume between the two groups. Additionally, the offspring in the AKW group and TPW group were given AKW and TPW, respectively, at weaning, and unexpectedly, the necrotic foci in the cardiac muscle were observed at the 15-week-old age in the AKW group, but not in the TPW group. The present study was designed to clarify the factors which are involved in that unusual increase of body weight and occurrence of cardiac necrosis. Eight dams in each group were given AKW or TPW (control) from day 0 of gestation to day 14 of lactation. The milk samples were collected on day 14 of lactation and analyzed for concentrations of calcium (Ca), sodium (Na), potassium (K), magnesium (Mg) and chloride (Cl). The AKW and TPW were also analyzed. Ca, Na and K levels in milk were significantly higher in the AKW group compared to the TPW group. No significant difference was noted in the Mg and Cl levels between the two groups. These data suggested that the Ca cation of AKW enriched the Ca concentration of the milk and accelerated the postnatal growth of the offspring of rats given AKW.

Wei Sheng Yan Jiu. 2004 Jul;33(4):422-5.

Impact of extra waters on immunosystem in mice

Li Y, Han C, Li Y, Li Y, Zhao X, Zhong K, Chen T, Zhang M, Fan F, Tao Y, Ji R.

Institute of Nutrition and Food Safety, Chinese Center for Disease Control and Prevention, Beijing 100021, China.

OBJECTIVE: To study impact of extra waters on immunosystem in mice-alkaline ionized water, mineral-ecology water, activated water, and pure water. METHODS: According to Function Assessment and Experiment for Function Food, 1996, the ratio of spain and body weight, the ratio of thymus and body weight, the delayed type hypersensitivity (DTH), the phagocytosing functions by cock RBC, the plaque forming cell (PFC) and HC50 testing were assessment with 120 male Balb/c mice of 17.8 - 23.3 g (group I), the mice were divided into four group, and drank daily the four kinds of waters respectively until 50 days. The lymph cell transformation by ConA and NK cell activity were assessment with other 120 male Balb/c mice of 17.8 - 23.3 g (group II), the mice were divided into four group and drank daily the four kinds of waters respectively until 50 days. The phagocytosing functions by carbon powder were assessment with other 120 female Balb/c mice of 16.7 - 22.0 g (group III) were divided into four group and drank daily the four kinds of waters respectively until 50 days. The data were statisticed by Stata soft. RESULTS: Other three waters compared with the pure water: (1) Alkaline ionized water and activated water can alleviate the body weight increase of male Balb/c mouse ($P < 0.01$ & $P < 0.05$), but alkaline ionized water, mineral-ecology water, activated water don't impact on the female Balb/c mouse body weight ($P > 0.05$). (2) Activated water can remarkably increase the ratio of thymus and body weight ($P < 0.05$), and increase the phagocytosing ability by cock RBC ($P < 0.01$), and increase the NK cell activity ($P < 0.01$). (3) The three extra waters don't impact on other items for the Balh/c mouce. CONCLUSION: The study must be continued to impact on immunosystem in mice for extra waters.

J Toxicol Sci. 1998 Dec;23(5):411-7.

Histopathological influence of alkaline ionized water on myocardial muscle of mother rats. Watanabe T, Shirai W, Pan I, Fukuda Y, Murasugi E, Sato T, Kamata H, Uwatoko K. Department of Veterinary Physiological Chemistry, College of Bioresource Sciences, Nihon University, Kanagawa, Japan.

We have reported that a marked necrosis and subsequent fibrosis of myocardium occurred among male rats 15 weeks old given alkaline ionized water (AKW) during gestation and suckling periods, and after weaning. In this study, it was examined whether similar lesions would occur in mother rats which were given AKW from day zero of gestation to day 20 of lactation. The myocardial lesion in the mother rats given AKW showed cell infiltration, vacuolation and fibrosis in the papillary muscle of the left ventricle, as were observed in male rats of 15 weeks old. Myocardial degeneration may cause a leakage of potassium into the blood that results in a higher concentration of potassium in the blood in the test group than in that of the control group given tap water.

J Toxicol Sci. 1998 Dec;23(5):365-71.

Influences of alkaline ionized water on milk yield, body weight of offspring and perinatal dam in rats. Watanabe T, Pan I, Fukuda Y, Murasugi E, Kamata H, Uwatoko K. Department of Veterinary Physiological Chemistry, College of Bioresource Sciences, Nihon University, Kanagawa, Japan.

The authors previously reported that male offspring of mothers rats given alkaline ionized water (AKW) showed a significantly higher body weight by day 14 after birth than did offspring of mother rats given tap water (TPW); furthermore, marked myocardial necrosis and fibrosis were observed particularly in the former male offspring at the age of 15 weeks. In the present experiment we looked for differences in bioparameters, namely the milk yield of mothers and suckled milk volume of the offspring, between the AKW- and the TPW-treated groups in order to reveal the factors which cause the unusual body weight gain in the offspring. Even though we were able to repeat our previous observation (the body weight of the male offspring of the AKW group increased significantly more by day 14 and 20 after birth and of the female by day 20 after birth than did that of the TPW group ($p < 0.05$), no significant difference was noted in any of the bioparameters, including those related to milk production and consumption. It is thus suspected that the water-hydrated cation, which was transferred either to the fetus through the placenta or to the offspring through the milk, might be the cause of the unusual body weight increase. Since calcium plays an important role in skeletal formation, it is tentatively concluded that the higher calcium concentration of AKW enriched the mother, serum calcium which was transferred to the fetus through the placenta and to the offspring through the milk.

J Toxicol Sci. 1995 May;20(2):135-42.

Effect of alkaline ionized water on reproduction in gestational and lactational rats. Watanabe T. Department of Veterinary Physiological Chemistry, College of Agriculture and Veterinary Medicine, Nihon University, Kanagawa, Japan.

Alkaline ionized water (AKW) produced by electrolysis was given to gestational and lactational rats, and its effect on dams, growth of fetuses and offsprings were investigated. The results showed that the intake of food and water in dams increased significantly when AKW was given from the latter half of the gestation period and from the former half of the lactation period. Body weight of the offsprings in the test group, both males and females, increased significantly from the latter half of the lactation period. During the lactation period and after weaning, the offsprings in the test group showed significantly hastened appearance of abdominal hair, eruption of upper incisors, opening of eyelids and other postnatal morphological developments both in males and females, as well as earlier separation of auricle and descent of testes in males compared with the control was noted. As mentioned above, it was suggested from the observations conducted that the AKW has substantial biological effects on postnatal growth, since intake of food and water and body weight of the offsprings increased and postnatal morphological development was also accelerated.

Adequate fluid replacement helps maintain hydration and, promotes the health, safety, and optimal physical performance of individuals participating in regular physical activity.
Medical Science Sports Exercise. 1996 Jan;28(1):i-vii. American College of Sports Medicine position stand. Exercise and fluid replacement. Convertino VA, Armstrong LE, Coyle EF, Mack GW, Sawka MN, Senay LC Jr, Sherman WM.

It is the position of the American College of Sports Medicine that adequate fluid replacement helps maintain hydration and, therefore, promotes the health, safety, and optimal physical performance of individuals participating in regular physical activity. This position statement is based on a comprehensive review and interpretation of scientific literature concerning the influence of fluid replacement on exercise performance and the risk of thermal injury associated with dehydration and hyperthermia.
Based on available evidence, the American College of Sports Medicine makes the following general recommendations on the amount and composition of fluid that should be ingested in preparation for, during, and after exercise or athletic competition:
1) It is recommended that individuals consume a nutritionally balanced diet and drink adequate fluids during the 24-hr period before an event, especially during the

period that includes the meal prior to exercise, to promote proper hydration before exercise or competition.
2) It is recommended that individuals drink about 500 ml (about 17 ounces) of fluid about 2 h before exercise to promote adequate hydration and allow time for excretion of excess ingested water.
3) During exercise, athletes should start drinking early and at regular intervals in an attempt to consume fluids at a rate sufficient to replace all the water lost through sweating (i.e., body weight loss), or consume the maximal amount that can be tolerated.
4) It is recommended that ingested fluids be cooler than ambient temperature [between 15 degrees and 22 degrees C (59 degrees and 72 degrees F])] and flavored to enhance palatability and promote fluid replacement. Fluids should be readily available and served in containers that allow adequate volumes to be ingested with ease and with minimal interruption of exercise.
5) Addition of proper amounts of carbohydrates and/or electrolytes to a fluid replacement solution is recommended for exercise events of duration greater than 1 h since it does not significantly impair water delivery to the body and may enhance performance. During exercise lasting less than 1 h, there is little evidence of physiological or physical performance differences between consuming a carbohydrate-electrolyte drink and plain water.
6) During intense exercise lasting longer than 1 h, it is recommended that carbohydrates be ingested at a rate of 30-60 g.h(-1) to maintain oxidation of carbohydrates and delay fatigue. This rate of carbohydrate intake can be achieved without compromising fluid delivery by drinking 600-1200 ml.h(-1) of solutions containing 4%-8% carbohydrates (g.100 ml(-1)). The carbohydrates can be sugars (glucose or sucrose) or starch (e.g., maltodextrin).
7) Inclusion of sodium (0.5-0.7 g.1(-1) of water) in the rehydration solution ingested during exercise lasting longer than 1 h is recommended since it may be advantageous in enhancing palatability, promoting fluid retention, and possibly preventing hyponatremia in certain individuals who drink excessive quantities of fluid. There is little physiological basis for the presence of sodium in oral rehydration solution for enhancing intestinal water absorption as long as sodium is sufficiently available from the previous meal.

Electrolyzed-reduced water scavenges active oxygen species and protects DNA from oxidative damage

Biochemical Biophysical Res Communication. 997 May 8;234(1):269-74.
Shirahata S , Kabayama S, Nakano M, Miura T, Kusumoto K, Gotoh M, Hayashi H , Otsubo K, Morisawa S, Katakura Y. Institute of Cellular Regulation Technology, Graduate School of Genetic Resources Technology, Kyushu University , Fukuoka , Japan . sirahata@grt.kyushu-u.ac.jp
Active oxygen species or free radicals are considered to cause extensive oxidative damage to biological macromolecules, which brings about a variety of diseases as well as aging. The ideal scavenger for active oxygen should be 'active hydrogen'. 'Active hydrogen' can be produced in reduced water near the cathode during

electrolysis of water. Reduced water exhibits high pH, low dissolved oxygen (DO), extremely high dissolved molecular hydrogen (DH), and extremely negative redox potential (RP) values. Strongly electrolyzed-reduced water, as well as ascorbic acid, (+)-catechin and tannic acid, completely scavenged O.-2 produced by the hypoxanthine-xanthine oxidase (HX-XOD) system in sodium phosphate buffer (pH 7.0).

The superoxide dismutase (SOD)-like activity of reduced water is stable at 4 degrees C for over a month and was not lost even after neutralization, repeated freezing and melting, deflation with sonication, vigorous mixing, boiling, repeated filtration, or closed autoclaving, but was lost by opened autoclaving or by closed autoclaving in the presence of tungsten trioxide which efficiently adsorbs active atomic hydrogen. Water bubbled with hydrogen gas exhibited low DO, extremely high DH and extremely low RP values, as does reduced water, but it has no SOD-like activity. These results suggest that the SOD-like activity of reduced water is not due to the dissolved molecular hydrogen but due to the dissolved atomic hydrogen (active hydrogen). Although SOD accumulated H2O2 when added to the HX-XOD system, reduced water decreased the amount of H2O2 produced by XOD. Reduced water, as well as catalase and ascorbic acid, could directly scavenge H2O2. Reduced water suppresses single-strand breakage of DNA b active oxygen species produced by the Cu(II)-catalyzed oxidation of ascorbic acid in a dose-dependent manner, suggesting that reduced water can scavenge not only O2.- and H2O2, but also 1O2 and .OH.

Comparison of electrolyzed oxidizing water with various antimicrobial interventions to reduce Salmonella species on poultry.

Poultry Science. 2002 Oct;81(10):1598-605. Fabrizio KA, Sharma RR, Demirci A, Cutter CN. Department of Food Science, The Pennsylvania State University , University Park 16802 , USA .

Foodborne pathogens in cell suspensions or attached to surfaces can be reduced by electrolyzed oxidizing (EO) water; however, the use of EO water against pathogens associated with poultry has not been explored. In this study, acidic EO water [EO-A; pH 2.6, chlorine (CL) 20 to 50 ppm, and oxidation-reduction potential (ORP) of 1,150 mV], basic EO water (EO-B; pH 11.6, ORP of -795 mV), CL, ozonated water (OZ), acetic acid (AA), or trisodium phosphate (TSP) was applied to broiler carcasses inoculated with Salmonella Typhimurium (ST) and submerged (4 C, 45 min), spray-washed (85 psi, 25 C, 15 s), or subjected to multiple interventions (EO-B spray, immersed in EO-A; AA or TSP spray, immersed in CL). Remaining bacterial populations were determined and compared at Day 0 and 7 of aerobic, refrigerated storage. At Day 0, submersion in TSP and AA reduced ST 1.41 log10, whereas EO-A water reduced ST approximately 0.86 log10. After 7 d of storage, EO-A water, OZ, TSP, and AA reduced ST, with detection only after selective enrichment.

Spray-washing treatments with any of the compounds did not reduce ST at Day 0. After 7 d of storage, TSP, AA, and EO-A water reduced ST 2.17, 2.31, and 1.06 log10, respectively. ST was reduced 2.11 log10 immediately following the multiple interventions, 3.81 log10 after 7 d of storage. Although effective against ST, TSP and AA are costly and adversely affect the environment. This study demonstrates that EO water can reduce ST on poultry surfaces following extended refrigerated storage.

Inactivation of Escherichia coli (O157:H7) and Listeria monocytogenes on plastic kitchen cutting boards by electrolyzed oxidizing water

Venkitanarayanan KS , Ezeike GO, Hung YC, Doyle MP. Department of Animal Science, University of Connecticut , Storrs 06269 , USA.

One milliliter of culture containing a five-strain mixture of Escherichia coli O157:H7 (approximately 10(10) CFU) was inoculated on a 100-cm2 area marked on unscarred cutting boards. Following inoculation, the boards were air-dried under a laminar flow hood for 1 h, immersed in 2 liters of electrolyzed oxidizing water or sterile de-ionized water at 23 degrees C or 35 degrees C for 10 or 20 min; 45 degrees C for 5 or 10 min; or 55 degrees C for 5 min. After each temperature-time combination, the surviving population of the pathogen on cutting boards and in soaking water was determined.

Soaking of inoculated cutting boards in electrolyzed oxidizing water reduced E. coli O157:H7 populations by > or = 5.0 log CFU/100 cm2 on cutting boards. However, immersion of cutting boards in de-ionized water decreased the pathogen count only by 1.0 to 1.5 log CFU/100 cm2. Treatment of cutting boards inoculated with Listeria monocytogenes in electrolyzed oxidizing water at selected temperature-time combinations (23 degrees C for 20 min, 35 degrees C for 10 min, and 45 degrees C for 10 min) substantially reduced the populations of L. monocytogenes in comparison to the counts recovered from the boards immersed in de-ionized water. E. coli O157:H7 and L. monocytogenes were not detected in electrolyzed oxidizing water after soaking treatment, whereas the pathogens survived in the de-ionized water used for soaking the cutting boards. This study revealed that immersion of kitchen cutting boards in electrolyzed oxidizing water could be used as an effective method for inactivating food-borne pathogens on smooth, plastic cutting boards.

PMID: 10456736 [Pub Med - indexed for MEDLINE]

The bactericidal effects of electrolyzed oxidizing water on bacterial strains involved in hospital infections

Vorobjeva NV , Vorobjeva LI, Khodjaev EY. Artificial Organs 2004 Ju n;28(6):590-2. Department of Physiology of Microorganisms, Biology Faculty, Moscow State University, Lenin Hills 1/12, Moscow 119992, Russia. nvvorobjeva@mail.ru

The study is designed to investigate bactericidal actions of electrolyzed oxidizing water on hospital infections. Ten of the most common opportunistic pathogens are used for this study. Cultures are inoculated in 4.5 mL of electrolyzed oxidizing (EO) water or 4.5 mL of sterile de-ionized water (control), and incubated for 0,

0.5, and 5 min at room temperature. At the exposure time of 30 s the EO water completely inactivates all of the bacterial strains, with the exception of vegetative cells and spores of bacilli which need 5 min to be killed. The results indicate that electrolyzed oxidizing water may be a useful disinfectant for hospital infections, but its clinical application has still to be evaluated.
PMID: 15153153 [Pub Med - in process]

Effect of electrolyzed oxidizing water and hydrocolloid occlusive dressings on excised burn-wounds in rats

Chin J Traumatol. 2003 Aug 1;6(4):234-7. Xin H, Zheng YJ, Hajime N, Han ZG.
Department of Thoracic Surgery , China - Japan Union Hospital , Jilin University, Jilin 130031, China . xinhua7254@yahoo.com.cn

OBJECTIVE: To study the efficacy of electrolyzed oxidizing water (EOW) and hydrocolloid occlusive dressings in the acceleration of epithelialization in excised burn-wounds in rats.

METHODS: Each of the anesthetized Sprague-Dawley rats (n=28) was subjected to a third-degree burn that covered approximately 10% of the total body surface area. Rats were assigned into four groups: Group I (no irrigation), Group II (irrigation with physiologic saline), Group III (irrigation with EOW) and Group IV (hydrocolloid occlusive dressing after EOW irrigation). Wounds were observed macroscopically until complete epithelialization was present, then the epithelialized wounds were examined microscopically.

RESULTS: Healing of the burn wounds was the fastest in Group IV treated with hydrocolloid occlusive dressing together with EOW. Although extensive regenerative epidermis was seen in each Group, the proliferations of lymphocytes and macrophages associated with dense collagen deposition were more extensive in Group II, III and IV than in Group I. These findings were particularly evident in Group III and IV.

CONCLUSIONS: Wound Healing may be accelerated by applying a hydrocolloid occlusive dressing on burn surfaces after they are cleaned with EOW.
PMID: 12857518 [Pub Med - indexed for MEDLINE]

Use of Ionized Water in hypochlorhydria or achlorhydria

Prof. Kuninaka Hironage, Head of Kuninaka Hospital

"Too many fats in the diets, which lead to the deposition of cholesterol on the blood vessels, which in turn constrict the blood flow, cause most illnesses such as high blood pressure. In accordance with the theory of Professor Gato of Kyushu University on Vitamin K (because vitamin K enables the blood calcium to increase) , or the consumption of more antioxidant water, the effectiveness of the increase in the calcium in high blood pressure is most significant. The consumption of alkaline antioxidant water for a period of 2 to 3 months, I have observed the blood pressure slowly drop, due to the water's solvent ability, which dissolves the cholesterol in the blood vessels ."

Effect of electrolyzed water on wound healing
Artif Organs. 2000 Dec;24(12):984-7. Yahagi N, Kono M, Kitahara M, Ohmura A, Sumita O, Hashimoto T, Hori K, Ning-Juan C, Woodson P, Kubota S, Murakami A, Takamoto S. Department of Anesthesiology, Teikyo University Mizonokuchi Hospital , Tokyo , Japan . naokiyah@aol.com
Electrolyzed water accelerated the healing of full-thickness cutaneous wounds in rats, but only anode chamber water (acid pH or neutralized) was effective. Hypochlorous acid (HOCl), also produced by electrolysis, was ineffective, suggesting that these types of electrolyzed water enhance wound healing by a mechanism unrelated to the well-known antibacterial action of HOCl. One possibility is that reactive oxygen species, shown to be electron spin resonance spectra present in anode chamber water, might trigger early wound healing through fibroblast migration and proliferation.
PMID: 11121980 [Pub Med - indexed for MEDLINE]

Physiological effects of alkaline Ionized Water: Effects on metabolites produced by intestinal fermentation by Takashi Hayakawa, Chicko Tushiya, Hisanori Onoda, Hisayo Ohkouchi, Harul-~to Tsuge (Gifu University, Faculty of Engineering, Dept. of Food Science)
We have found that long-term ingestion of alkaline ionized water (AIW) reduces cecal[92] fermentation in rats that were given highly fermentable commercial diet (MF: Oriental Yeast Co., Ltd.). In this experiment, rats were fed MF and test water (tap water, AIW with pH at 9 and 10) for about 3 months. Feces were collected on the 57th day, and the rats were dissected on the 88th day. The amount of ammonium in fresh feces and cecal contents as well as fecal free-glucose tended to drop down for the AIW group. In most cases, the amount of free-amino acids in cecal contents did not differ significantly except for cysteine (decreased in AIW with pH at 10) and isoleucine (increased in AIW with pH at 10).
Purpose of tests
Alkaline ionized water electrolyzers have been approved for manufacturing in 1965 by the Ministry of Health and Welfare as medical equipment to produce medical substances. Alkaline ionized water (AIW) produced by this equipment is known to be effective against gastrointestinal fermentation, chronic diarrhea, indigestion and hyperchylia as well as for controlling gastric acid.*1 This is mainly based on efficacy of the official calcium hydroxide. *2 By giving AIW to rats for a comparatively long time under the condition of extremely high level of intestinal fermentation, we have demonstrated that AIW intake is effective for inhibition of intestinal fermentation when its level is high based on some test results where AIW worked against cecal hypertrophy and for reduction in the amount of short-chain fatty acid that is the main product of fermentation.*3 We have reported that this is caused by the synergy between calcium level generally contained in AIW (about 50ppm) and the value of pH, and that frequency of detecting some anaerobic bacteria tends to be higher in alkaline Ionized Water groups than the other, although the bacteria count in the intestine does not have significant difference.

Based on these results, we made a judgment that effect of taking AIW supports part of inhibition mechanism against abnormal intestinal fermentation, which is one of the claims of efficacy that have been attributed to alkaline ionized water electrolyzers. *4 On the other hand, under the dietary condition of low intestinal fermentation, AIW uptake does not seem to inhibit fermentation that leads us to believe that effect of AIW uptake is characteristic of hyper-fermentation state. Metabolites produced by intestinal fermentation include indole and skatole in addition to organic acids such as short-chain fatty acid and lactic acid as well as toxic metabolites such as ammonium, phenol and pcresol. We do not know how AIW uptake would affect the production of these materials. In this experiment, we have tested on ammonium production as explained in the following sections.

Testing methods

Four-week-old male Wistar/ST Clean rats were purchased from Japan SLC Co., Ltd. and were divided into 3 groups of 8 each after preliminary breeding. AIW of pH 9 and 10 was produced by an electrolyzer Mineone ROYAL NDX3 1 OH by Omco Co., Ltd. This model produces AIW by electrolyzing water with calcium lactate added. On the last day of testing, the rats were dissected under Nembutal anesthesia to take blood from the heart by a heparin-treated syringe. As to their organs, the small intestines, cecum and colon plus rectum were taken out from each of them. The cecurn was weighed and cleaned with physiological saline after its contents were removed, and the tissue weight was measured after wiping out moisture. Part of cecal contents was measured its pH, and the rest was used to assay ammonium concentration. The amount of ammonium contained in fresh feces and cecal contents was measured by the Nessler method after collecting it in the extracted samples using Conway 's micro-diffusion container. Fecal free-glucose was assayed by the oxygen method after extraction by hot water. Analysis of free amino acids contained in cecal contents was conducted by the Waters PicoTag amino acid analysis system.

Test results and analyses

No difference was found in the rats' weight gain, water and feed intake and feeding efficiency, nor was any particular distinction in appearance identified. The length of the small intestines and colon plus rectum tended to decline in AIW groups. PH value of cecal contents was higher and the amount of fecal free-glucose tended to be lower in AIW groups than the control group. Since there was no difference in fecal discharge itself, the amount of free-glucose discharged per day was at a low level. The amount of discharged free-glucose in feces is greater when intestinal fermentation is more intensive, which indicates that intestinal fermentation is more inhibited in AIW groups than the control group. Ammonium concentration in cecal contents tends to drop down in AIW groups (Fig. 1). This trend was most distinctive in case of fresh feces of one of AIW groups with pH 10 (Fig.2) AIW uptake was found to be inhibitory against ammonium production. In order to study dynamics of amino acids in large intestines, we examined free amino acids in the cecal contents to find out that cysteine level is low in AIW groups whereas isoleucine level is high in one of AIW groups with pH 10, although no significant difference was identified for other amino acids.

References

"Verification of Alkaline Ionized Water" by Life Water Institute, Metamor Publishing Co., 1994, p.46
"Official Pharmaceutical Guidelines of Japan , Vol. IT' by Japan Public Documents Association, Hirokawa Publishing Co. , 1996
"Science and Technology of Functional Water" (part) by Takashi Hayakawa, Haruffito Tsuge, edited by Water Scienll cc Institute, 1999, pp.109-116
"Tasics and Effective Use of Alkaline Ionized Water" by Takashi Hayakawa, Haruhito Tsuge, edited by Tetsuji Hc kudou, 25th General Assembly of Japan Medical Congress Functional Water in Medical Treatment", Administration Offices, 1999, pp. 10- 11

Clinical Improvements Obtained From The Intake Of Reduced Water

Extracts from “Presentation At The Eight Annual International Symposium On man And His Environment in Health And Disease" on February 24th 1990, at The Grand Kempinski Hotel, Dallas, Texas, USA by Dr. H. Hayashi, M.D. and Dr. M Kawamura, M.D.,

Since the introduction of alkaline ionic water in our clinic in 1985, we have had the following interesting clinical experiences in the use of this type of water. By the use of alkaline ionic water for drinking and the preparation of meals for our in-patients, we have noticed :

- Declines in blood sugar levels in diabetic patients.
- Improvements in peripheral circulation in diabetic gangrene.
- Declines in uric acid levels in patients with gout.
- Improvements in liver function exams in hepatic disorders.
- Improvements in gastroduodenal ulcer and prevention of their recurrences.
- Improvements in hypertension and hypotension.
- Improvements in allergic disorders such as asthma, urticaria, rhinites and atopic dermatitis.
- Improvements in persistent diarrhea which occurred after gastrectomy.
- Quicker improvements in post operative bower paralysis.
- Improvements in serum bilirubin levels in new born babies.

Being confirming clinical improvements, we have always observed changes of stools of the patients, with the color of their feces changing from black-brown color to a brighter yellow-brown one, and the odor of their feces becoming almost negligible.

The number of patients complaining of constipation also decreased markedly. The change of stool findings strongly suggests that alkaline ionic water intake can decrease the production of putrefied or pathogenic metabolites.
Devices to produce reduced water were introduced into our clinic in May 1985. Based on the clinical experiences obtained in the past 15 years, it can be said that introduction of electrolyzed-reduced water for drinking and cooking purpose for in-patients should be the very prerequisite in our daily medical practices. Any

dietary recipe cannot be a scientific one if property of water is not taken by the patients is not taken into consideration.
The Ministry of Health and Welfare in Japan announced in 1965 that the intake of reduced water is effective for restoration of intestinal flora metabolism.

Decoposition of ethylene, a flower-senescence hormone, with electrolyzed anode water. Biosci Biotechnology Biochemistry 2003 Apr;67(4):790-6. Harada K, Yasui K. Department of Research and Development, Hokkaido Electric Power Co., Inc., 2-1 Tsuishikari, Ebetsu, Hokkaido 067-0033, Japan.
kharada@h1.hotcn.ne.jp

Electrolyzed anode water (EAW) markedly extended the vase life of cut carnation flowers. Therefore, a flower-senescence hormone involving ethylene decomposition by EAW with potassium chloride as an electrolyte was investigated. Ethylene was added externally to EAW, and the reaction between ethylen and the available chlorine in EAW was examined. EAW had a low pH value (2.5), a high concentration of dissolved oxygen, and extremely high redox potential (19.2 mg/l and 1323 mV, respectively) when available chlorine was at a concentration of about 620 microns.

The addition of ethylene to EAW led to ethylene decomposition, and an equimolar amount of ethylene chlorohydrine with available chlorine was produced. The ethylene chlorohydrine production was greatly affected by the pH value (pH 2.5, 5.0 and 10.0 were tested), and was faster in an acidic solution. Ethylene chlorohydrine was not produced after ethylene had been added to EAW at pH 2.6 when available chlorine was absent, but was produced after potassium hypochlorite had been added to such EAW. The effect of the pH value of EAW on the vase life of cut carnations was compatible with the decomposition rate of ethylene in EAW of the same pH value. These results suggest that the effect of EAW on the vase life of cut carnations was due to the decomposition of ethylene to ethylene chlorohydrine by chlorine from chlorine compounds.
PMID: 12784619 [Pub Med - indexed for MEDLINE]

Use of *Ionized Water* in treating Acidosis
Prof. Hatori Tasutaroo, Head of Akajiuiji Blood Centre, Yokohama Hospital , Faitama District
" Due to a higher standard of living, our eating habits have changed. We consume too much proteins, fats and sugar. The excess fats and carbohydrates are in the body as fats. In the present lifestyles, Americans are more extravagant on food compared to the Japanese. Due to this excessive intake obesity is a significant problem. Normally, one out of five males and one out of four females is obese. The degree of "burn-out" in food intake largely depends on the amount on intake of vitamins and minerals. When excessive intake of proteins, carbohydrates and fats occurs, the requirement for vitamins and minerals increases. However, there is not much research carried out pertaining to the importance of vitamins and minerals.
Nowadays, many people suffer from acidification that leads to diabetes, heart

diseases, cancer, live and kidney diseases. If our food intake can be completely burned off, then there is no deposition of fats. Obviously, there will be no acidification problem and hence there should not be any sign of obesity.
The antioxidant water contains an abundance of ionic calcium. This ionic calcium helps in the "burn-off" process. By drinking antioxidant water, it provides sufficient minerals for our body. As a result, we do not need to watch our diet to stay slim.
Hence, antioxidant water is a savior for those suffering from obesity and many adult diseases, providing good assistance in enhancing good health."

Allergies and *Ionized Water*
Prof. Kuninaka Hironaga, Head of Kuninaka Hospital
"Mr. Yamada, the head of Police Research Institute, suffered from severe allergy. He was treated repeatedly by skin specialist, but with no success. Then he started consuming antioxidant water. The allergy responded very well and was soon completely cured. No relapse had occurred, although he had taken all kinds of food. He was most grateful and excited about this treatment.
As for myself, I had also suffered severe allergy. Ever since I began to consume antioxidant water, the allergy has recovered. Since then, I started a research on the effectiveness of antioxidant water.
I discovered that most allergies are due to acidification of body condition and is also related to consuming too much meat and sugar. In every allergy case, the patient's antioxidant minerals are excessively low which in turn lower the body resistance significantly. The body becomes overly sensitive and develops allergy easily. To stabilize the sensitivity, calcium solution in injected into the vein. Therefore, it is clear that the antioxidant water has ionic calcium, which can help alleviate allergy.
The ionic calcium not only enhances the heart, urination, and neutralization of toxins but controls acidity. It also enhances the digestive system and liver function. This will promote natural healing power and hence increase its resistance to allergy. In some special cases of illness, which do not respond to drugs, it is found, it is found to respond well to antioxidant water."

Use of *Ionized Water* for gynecological conditions and treatments
Prof. Watanabe Ifao, Watanabe Hospital
Ionized alkaline antioxidant water improves body constituents and ensures effective healing to many illnesses. The uses of antioxidant water in gynecological patients have proved to be very effective. The main reason for its effectiveness is that this water can neutralize toxins.
When given antioxidant water to pre-eclamptic toxemia cases, the results are most significant. During my long years of servicing the pre-eclamptic toxemia cases, I found that the women with pre-eclamptic toxemia who consumed antioxidant

water tend to deliver healthier babies with stronger muscles . A survey report carried out on babies in this group showed intelligence above average.

The mechanism of the enhanced antioxidant effects against superoxide anion radicals of reduced water produced by electrolysis

Biophysical Chemistry. 2004 Jan 1;107(1):71-82. Hanaoka K, Sun D, Lawrence R, Kamitani Y, Fernandes G. Bio-REDOX Laboratory Inc. 1187-4, Oaza-Ueda, Ueda-shi, Nagano-ken 386-0001, Japan . hanak@rapid.ocn.ne.jp

We reported that reduced water produced by electrolysis enhanced the antioxidant effects of proton donors such as ascorbic acid (AsA) in a previous paper. We also demonstrated that reduced water produced by electrolysis of 2 mM NaCl solutions did not show antioxidant effects by itself. We reasoned that the enhancement of antioxidant effects may be due to the increase of the ionic product of water as solvent. The ionic product of water (pKw) was estimated by measurements of pH and by a neutralization titration method. As an indicator of oxidative damage, Reactive Oxygen Species- (ROS) mediated DNA strand breaks were measured by the conversion of supercoiled phiX-174 RF I double-strand DNA to open and linear forms. Reduced water had a tendency to suppress single-strand breakage of DNA induced by reactive oxygen species produced by H2O2/Cu (II) and HQ/Cu (II) systems. The enhancement of superoxide anion radical dismutation activity can be explained by changes in the ionic product of water in the reduced water.

Toxin Neutralization with the Use of Electrolysed Water

Prof. Kuwata Keijiroo, Doctor of Medicine

" In my opinion, the wonder of antioxidant water is the ability neutralizes toxins, but it is not a medicine. The difference is that the medicine can only apply to each and individual case, whereas the antioxidant water can be consumed generally and its neutralizing power is something which is very much unexpected. Now, in brief, let me introduce to you a heart disease case and how it was cured.

The patient was a 35 years old male suffering from vascular heart disease. For 5 years, his sickness deteriorated. He was in the Setagays Government Hospital for treatment.

During those 5 years, he had been in and out of the hospital 5 to 6 times. He had undergone high tech examinations such as angiogram by injecting VINYL via the vein into the heart. He consulted and sought treatment from many good doctors where later he underwent a major surgical operation. Upon his discharge from the hospital, he quit his job to convalesce. However, each time when his illness relapsed, the attack seemed to be even more severe.

Last year, in August, his relatives were in despair and expected he would not live much longer. It so happened at that time that the victim's relative came across antioxidant water processor. His illness responded well and he is now on the road to recovery. "

In the United States , cardiovascular diseases account for more than one-half of the approximate 2 million deaths occurring each year.... It is estimated that optimal conditioning of drinking water could reduce this cardiovascular disease mortality

rate by as much as 15 percent in the United States
Report of the Safe Drinking Water Committee of the National Academy of Sciences, 1977

Eczema and the Effects of *Ionized Water*
Prof. Tamura Tatsuji, Keifuku Rehabilitation Center
"Eczema is used to describe several varieties of skin conditions, which have a number of common features. The exact cause or causes of eczema are not fully understood. I many cases, eczema can be attributed by external irritants.
Let me introduce a patient who recovered from skin disease after consuming the antioxidant water. This patient suffered 10 years of eczema and could not be cured effectively even under specialist treatment. This patient, who is 70 years of age, is the president of a vehicle spare parts company. After the war, his lower limbs suffered acute eczema, which later became chronic. He was repeatedly treated in a specialist skin hospital.
The left limb responded well to treatment, but not so on the right limb. He suffered severe itchiness, which, when scratched led to bleeding. During the last 10 years, he was seen and treated by many doctors. When I first examined him, his lower limb around the joints was covered with vesicles. Weeping occurred owing to serum exuding from the vesicles.
I advised him to try consuming antioxidant water. He bought a unit and consumed the antioxidant water religiously and used the acidic water to bathe the affected areas. After 2 weeks of treatment the vesicles dried up. The eczema was completely cleared without any relapse after 1½ month."

Diabetes and the Effects of *Ionized Water*
Prof. Kuwata Keijiroo, Doctor of Medicine
"When I was serving in the Fire Insurance Association, I used to examine many diabetic patients. Besides treating them with drugs, I provided them with antioxidant water. After drinking antioxidant water for one month, 15 diabetic patients were selected and sent to Tokyo University for further test and observations.
Initially, the more serious patients were a bit apprehensive about the treatment. When the antioxidant water was consumed for some time, the sugar in the blood and urine ranged from a ratio of 300 mg/l to 2 mg / dc. There was a time where the patient had undergone 5 to 6 blood tests a day and detected to be within normal range. Results also showed that even 1 ½ hour after meals, the blood sugar and urine ratio was 100 mg/dc: 0 mg/dc . The sugar in the urine has completely disappeared."
NOTE: More Americans than ever before are suffering from diabetes, with the number of new cases averaging almost 800,000 each year. The disease has steadily increased in the United States since 1980, and in 1998, 16 million Americans were diagnosed with diabetes (10.3 million diagnosed; 5.4 million undiagnosed). Diabetes is the seventh leading cause of death in the United States , and more than 193,000 died from the disease and its related complication in 1996. ....

The greatest increase - 76 percent - occurred in people age 30 to 30. ....
From: U. S. Department of Health and Human Services, October 13, 2000 Fact Sheet.

Digestive Problems and Ionized Water

Prof. Kogure Keizou, Kogure Clinic of Juntendo Hospital

The stomach is readily upset both by diseases affecting the stomach and by other general illnesses. In addition, any nervous tension or anxiety frequently causes gastric upset, vague symptoms when This information is under some strain.

The important role of antioxidant water in our stomach is to neutralize the secretion and strengthen it s functions. Usually, after consuming the antioxidant water for 1 to 3 minutes, the gastric juice increase to 1½ times. For those suffering from hypochlorhydria or achlorhydria (low in gastric juice) the presence of antioxidant water will stimulate the stomach cells to secrete more gastric juice. This in turn enhances digestion and absorption of minerals.

However, on the other hand, those with hyperchlorhydria (high in gastric juice), the antioxidant water neutralizes the excessive gastric juice. Hence, it does not create any adverse reaction.

According to the medical lecturer from Maeba University , the pH of the gastric secretion will still remain normal when antioxidant water is consumed. This proves that the ability of the antioxidant water is able to neutralize as well as to stimulate the secretion."

Reduced Water for the Prevention of Disease

Dr.Sanetaka Shirahata

Graduate school of Genetic Resources Technology , Kyushu University ,

6-10-1 Hakozaki, Higashi-ku, Fukuoka 812-8581, Japan.

It has long been established that reactive oxygen species (ROS) cause many types of damage to biomolecules and cellular structures that, in turn result in the development of a variety of pathologic states such as diabetes, cancer and aging. Reduced water is defined as anti-oxidative water produced by reduction of water. Electrolyzed reduced water (ERW) has been demonstrated to be hydrogen-rich water and can scavenge ROS in vitro (Shirahata et al., 1997).

The reduction of proton in water to active hydrogen (atomic hydrogen, hydrogen radical) that can scavenge ROS is very easily caused by a weak current, compared to oxidation of hydroxyl ion to oxygen molecule. Activation of water by magnetic field, collision, minerals etc. will also produce reduced water containing active hydrogen and/or hydrogen molecule. Several natural waters such as Hita Tenryosui water drawn from deep underground in Hita city in Japan, Nordenau water in Germany and Tlacote water in Mexico are known to alleviate various diseases. We have developed a sensitive method by which we can detect active hydrogen existing in reduced water, and have demonstrated that not only ERW but also natural reduced waters described above contain active hydrogen and scavenge ROS in cultured cells. ROS is known to cause reduction of glucose uptake by inhibiting the insulin-signaling pathway in cultured cells. Reduced water scavenged intracellular ROS and stimulated glucose uptake in the presence or absence of insulin in both rat L6 skeletal muscle cells and mouse 3T3/L1 adipocytes. This insulin-like activity of reduced water was inhibited by wortmannin that is specific inhibitor of PI-3 kinase, a key molecule in insulin signaling pathways. Reduced water protected insulin-responsive cells from sugar toxicity and improved the damaged sugar tolerance of type 2 diabetes model mice, suggesting that reduced water may improve insulin-independent diabetes mellitus. Cancer cells are generally exposed to high oxidative stress. Reduced water cause impaired tumor phenotypes of human cancer cells, such as reduced growth rate, morphological changes, reduced colony formation ability in soft agar, passage number-dependent telomere shortening, reduced binding abilities of telomere binding proteins and suppressed metastasis. Reduced water suppressed the growth of cancer cells transplanted into mice, demonstrating their anti-cancer effects in vivo. Reduced water will be applicable to not only medicine but also food industries, agriculture, and manufacturing industries.

Shirahata, S. et al .: Electrolyzed reduced water scavenges active oxygen species and protects DNA from oxidative damage. Biochem. Biophys. Res. Commun., 234, 269174, 1997.

Clinical evaluation of alkaline *Ionized Water* for abdominal complaints: Placebo controlled double blind tests *by Hirokazu Tashiro, Tetsuji Hokudo, Hiromi Ono, Yoshihide Fujiyama, Tadao Baba (National Ohkura Hospital, Dept. of Gastroenterology; Institute of Clinical Research, Shiga University of Medical Science, Second Dept. of Internal Medicine)*
Effect of alkaline *Ionized Water* on abdominal complaints was evaluated by placebo controlled double blind tests. Overall scores of improvement using alkaline *Ionized Water* marked higher than those of placebo controlled group, and its effect proved to be significantly higher especially in slight symptoms of chronic diarrhea and abdominal complaints in cases of general malaise. Alkaline *Ionized Water* group did not get interrupted in the course of the test, nor did it show serious side effects nor abnormal test data. It was confirmed that alkaline *Ionized Water* is safer and more effective than placebos.

Summary
Effect of alkaline *Ionized Water* on abdominal complaints was clinically examined by double blind tests using clean water as placebo. Overall improvement rate was higher for alkaline *Ionized Water* group than placebo group and the former proved to be significantly more effective than the other especially in cases of slight symptoms. Examining improvement rate for each case of chronic diarrhea, constipation and abdominal complaints, alkaline *Ionized Water* group turned out to be more effective than placebo group for chronic diarrhea, and abdominal complaints. The test was stopped in one case of chronic diarrhea, among placebo group due to exacerbation, whereas alkaline *Ionized Water* group did not stop testing without serious side effects or abnormal test data in all cases. It was

confirmed that alkaline *Ionized Water* is more effective than clean water against chronic diarrhea, abdominal complaints and overall improvement rate (relief of abdominal complaints) and safer than clean water.

Introduction

Since the approval of alkaline *Ionized Water* electrolyzers by Pharmaceutical Affairs Law in 1966 for its antacid effect and efficacy against gastrointestinal disorders including hyperchylia, indigestion, abnormal gastrointestinal fermentation and chronic diarrhea, they have been extensively used among patients. However, medical and scientific evaluation of their validity is not established. In our study, we examined clinical effect of alkaline *Ionized Water* on gastrointestinal disorders across many symptoms in various facilities. Particularly, we studied safety and usefulness of alkaline *Ionized Water* by double-blind tests using clean water as a control group.

Test subjects and methods

163 patients (34 men, 129 women, age 21 to 72, average 38.6 years old) of indigestion, abnormal gastrointestinal fermentation (with abnormal gas emission and rugitus) and abdominal complaints caused by irregular dejection (chronic diarrhea, or constipation) were tested as subjects with good informed consent. Placebo controlled double blind tests were conducted using alkaline *Ionized Water* and clean water at multiple facilities. An alkaline *Ionized Water* electrolyzer sold commercially was installed with a pump driven calcium dispenser in each of the subject homes. Tested alkaline *Ionized Water* had pH at 9.5 and calcium concentration at 30ppm. Each subject in placebo group used a water purifier that has the same appearance as the electrolyzer and produces clean water.

The tested equipment was randomly assigned by a controller who scaled off the key code which was stored safely until the tests were completed and the seal was opened again.

Water samples were given to each patient in the amount of 200ml in the morning with the total of 50OmI or more per day for a month. Before and after the tests, blood, urine and stool were tested and a log was kept on the subjective symptoms, bowel movements and accessory symptoms. After the tests, the results were analyzed based on the log and the test data.

Test Results

1. Symptom. Among 163 tested subjects, alkaline *Ionized Water* group included 84 and placebo group 79. Background factors such as gender, age and basal disorders did not contribute to significant difference in the results.

2. Overall improvement rate. As to overall improvement rate of abdominal complaints, alkaline *Ionized Water* group had 2 cases of outstanding improvement (2.5%), 26 cases of fair improvement (32.1%), 36 cases of slight improvement (44.4%), 13 cases of no change (16%) and 4 cases of exacerbation (4.9%), whereas placebo group exhibited 4 (5.2%), 19 (24.7%), 27 (35.1%), 25 (32.5%) and 2 cases (2.6%) for the same category. Comparison between alkaline *Ionized Water* and placebo groups did not reveal any significant difference at the level of 5% significance according to the Wilcoxon test, although alkaline *Ionized Water* group

turned out to be significantly more effective than placebo group at the level of p value of 0.22.
3. Improvement rate by basal symptom. Basal symptoms were divided into chronic diarrhea, constipation and abdominal complaints (dyspepsia) and overall improvement rate was evaluated for each of them to study effect of alkaline *Ionized Water*. In case of chronic diarrhea, alkaline *Ionized Water* group resulted in 94.1% of effective cases and 5.9% of non effective cases. Placebo group came up with 64,7% effective and 35.3% non effective. These results indicate alkaline *Ionized Water* group proved to be significantly more effective than placebo group. In case of slighter chronic diarrhea, comparison between groups revealed that alkaline *Ionized Water* group is significantly more effective than placebo group ($p=0.015$). In case of constipation, alkaline *Ionized Water* group consisted of 80.5% of effective and 19.5% of non effective cases, whereas placebo group resulted in 73.3% effective and 26.3 non effective. As to abdominal complaints (dyspepsia), alkaline *Ionized Water* group had 85.7% of effective and 14.3% non effective cases while placebo group showed 47.1% and 62.9% respectively. Alkaline *Ionized Water* group proved to be significantly more effective than placebo group ($p=0.025$).

Conclusion

As a result of double blind clinical tests of alkaline *Ionized Water* and clean water, alkaline *Ionized Water* was proved to be more effective than clean water against chronic diarrhea, abdominal complaints (dyspepsia) and overall improvement rate (relief from abdominal complaints). Also, safety of alkaline *Ionized Water* was confirmed which clinically verifies its usefulness.

Treatment of Escherichia coli (O157:H7) inoculated alfalfa seeds and sprouts with electrolyzed oxidizing water

International Journal Food Microbiology 2003 Sep 15;86(3):231-7. Department of Agricultural and Biological Engineering, Pennsylvania State University , University Park , PA 16802 , USA .

Electrolyzed oxidizing water is a relatively new concept that has been utilized in agriculture, livestock management, medical sterilization, and food sanitation. Electrolyzed oxidizing (EO) water generated by passing sodium chloride solution through an EO water generator was used to treat alfalfa seeds and sprouts inoculated with a five-strain cocktail of nalidixic acid resistant Escherichia coli O157:H7. EO water had a pH of 2.6, an oxidation-reduction potential of 1150 mV and about 50 ppm free chlorine. The percentage reduction in bacterial load was determined for reaction times of 2, 4, 8, 16, 32, and 64 min. Mechanical agitation was done while treating the seeds at different time intervals to increase the effectiveness of the treatment. Since E. coli O157:H7 was released due to soaking during treatment, the initial counts on seeds and sprouts were determined by soaking the contaminated seeds/sprouts in 0.1% peptone water for a period equivalent to treatment time. The samples were then pummeled in 0.1% peptone water and spread plated on tryptic soy agar with 5 microg/ml of nalidixic acid

(TSAN). Results showed that there were reductions between 38.2% and 97.1% (0.22-1.56 log(10) CFU/g) in the bacterial load of treated seeds.
The reductions for sprouts were between 91.1% and 99.8% (1.05-2.72 log(10) CFU/g). An increase in treatment time increased the percentage reduction of E. coli O157:H7. However, germination of the treated seeds reduced from 92% to 49% as amperage to make EO water and soaking time increased. EO water did not cause any visible damage to the sprouts.
Ionized Acid Water promotes substantially healthier plant growth.

A clinical study of liver abscesses at the Critical Care and Emergency Center of Wate Medical University] Fujino Y, Inoue Y, Onodera M, Yaegashi Y, Sato N, Endo S, Omori H, Suzuki K. [Article in Japanese] Department of Critical Care Medicine, Wate Medical University.

We studied 13 emergency cases of liver abscess. Five cases of septic shock or clouding of consciousness were identified on admission. Six patients had diabetes mellitus. Twelve patients met the diagnostic criteria for systemic inflammatory response syndrome, and nine met the criteria for disseminated intravascular coagulation. Plasma endotoxin levels improved rapidly after drainage. Causative organisms were isolated in all patients, and the most common organism was Klebsiella pneumoniae (seven cases). Percutaneous transhepatic abscess drainage (PTAD) was performed not only in single cases but also in multiple cases with main huge abscesses. Surgical treatment was performed in the following three cases: a ruptured abscess, an ineffective PTAD, and a case of peritonitis after PTAD. Irrigation of abscesses with strong acidic electrolyzed water revealed a significant decrease in treatment duration. In the majority of our cases, severe conditions were identified on admission. Strong acidic electrolyzed water was useful for management of PTAD.

Evaluation of hypochlorous acid washes in the treatment of chronic venous leg ulcers. *J.B. Selkon, MBChB, FRCPath, DCP, Honorary Consultant Microbiologist, Oxford University, Department of Clinical Medicine; G.W. Cherry, DPhil (Oxon), Clinical Faculty Member, Oxford University Medical School; Chairman, Oxford International Wound Healing Foundation; J.M. Wilson, RN, Clinical Research Nurse; M.A. Hughes, PhD, Clinical Scientist, Oxford International Wound Healing Foundation; all at Dermatology Department, Churchill Hospital, Oxford, UK. Email: jselkon@onetel.com VOL 15, NO 1, Page 33 , January 2006*

Hypochlorous acid is a highly microbial agent active against bacteria, viruses and fungi. This study aimed to determine if it has a role as an additional treatment for chronic venous leg ulcers that have not healed with standard treatment. The patients acted as their own controls, in that only those who failed to achieve a 44% reduction in wound size with compression therapy after three weeks received the new treatment. Of the 30 patients recruited into the study, 20 were given the hypochlorous acid washes. In these patients, nine ulcers healed and five reduced in size by 60% in the follow-up period. All patients' ulcers became pain-free.

Enhancing the bactericidal effect of electrolyzed water on Listeria monocytogenes biofilms formed on stainless steel. Ayebah B, Hung YC, Frank JF. Department of Food Science and Technology, University of Georgia, 1109 Experiment Street, Griffin, Georgia 30223, USA.

Biofilms are potential sources of contamination to food in processing plants, because they frequently survive sanitizer treatments during cleaning. The objective of this research was to investigate the combined use of alkaline and acidic electrolyzed (EO) water in the inactivation of Listeria monocytogenes biofilms on stainless steel surfaces. Biofilms were grown on rectangular stainless steel (type 304, no. 4 finish) coupons (2 by 5 cm) in a 1:10 dilution of tryptic soy broth that contained a five-strain mixture of L. monocytogenes for 48 h at 25 degrees C. The coupons with biofilms were then treated with acidic EO water or alkaline EO water or with alkaline EO water followed by acidic EO water produced at 14 and 20 A for 30, 60, and 120 s. Alkaline EO water alone did not produce significant reductions in L. monocytogenes biofilms when compared with the control. Treatment with acidic EO water only for 30 to 120 s, on the other hand, reduced the viable bacterial populations in the biofilms by 4.3 to 5.2 log CFU per coupon, whereas the combined treatment of alkaline EO water followed by acidic EO water produced an additional 0.3- to 1.2-log CFU per coupon reduction. The population of L. monocytogenes reduced by treatments with acidic EO water increased significantly with increasing time of exposure. However, no significant differences occurred between treatments with EO water produced at 14 and 20 A. Results suggest that alkaline and acidic EO water can be used together to achieve a better inactivation of biofilms than when applied individually.

Selective stimulation of the growth of anaerobic microflora in the human intestinal tract by electrolyzed reducing water. Vorobjeva NV. Department of Physiology of Microorganisms, Biology Faculty, Lomonosov Moscow State University, 119992 Moscow, Russia. nvvorobjeva@mail.ru

96-99% of the "friendly" or residential microflora of intestinal tract of humans consists of strict anaerobes and only 1-4% of aerobes. Many diseases of the intestine are due to a disturbance in the balance of the microorganisms inhabiting the gut. The treatment of such diseases involves the restoration of the quantity and/or balance of residential microflora in the intestinal tract. It is known that aerobes and anaerobes grow at different oxidation-reduction potentials (ORP). The former require positive E(h) values up to +400 mV. Anaerobes do not grow unless the E(h) value is negative between -300 and -400 mV. In this work, it is suggested that prerequisite for the recovery and maintenance of obligatory anaerobic microflora in the intestinal tract is a negative ORP value of the intestinal milieu. Electrolyzed reducing water with E(h) values between 0 and -300 mV produced in electrolysis devices possesses this property. Drinking such water favors the growth of residential microflora in the gut. A sufficient array of data confirms this idea. However, most researchers explain the mechanism of its action by an antioxidant properties destined to detox the oxidants in the gut and other host tissues. Evidence is presented in favor of the hypothesis that the primary target for electrolyzed reducing water is the residential microflora in the gut.

Efficacy of electrolyzed water in inactivating Salmonella enteritidis and Listeria monocytogenes on shell eggs. Journal of Food Protection 2005 May;68(5):986-90. Park CM, Hung YC, Lin CS, Brackett RE. Department of Food Science and Technology, University of Georgia, Griffin, Georgia 30223-1797, USA.

The efficacy of acidic electrolyzed (EO) water produced at three levels of total available chlorine (16, 41, and 77 mg/ liter) and chlorinated water with 45 and 200 mg/liter of residual chlorine was investigated for inactivating Salmonella Enteritidis and Listeria monocytogenes on shell eggs. An increasing reduction in Listeria population was observed with increasing chlorine concentration from 16 to 77 mg/liter and treatment time from 1 to 5 min, resulting in a maximal reduction of 3.70 log CFU per shell egg compared with a de-ionized water wash for 5 min. There was no significant difference in antibacterial activities against Salmonella and Listeria at the same treatment time between 45 mg/liter of chlorinated water and 14-A acidic EO water treatment (P > or = 0.05). Chlorinated water (200 mg/liter) wash for 3 and 5 min was the most effective treatment; it reduced mean populations of Listeria and Salmonella on inoculated eggs by 4.89 and 3.83 log CFU/shell egg, respectively. However, reductions (log CFU/shell egg) of Listeria (4.39) and Salmonella (3.66) by 1-min alkaline EO water treatment followed by another 1 min of 14-A acidic EO water (41 mg/liter chlorine) treatment had a similar reduction to the 1-min 200 mg/liter chlorinated water treatment for Listeria (4.01) and Salmonella (3.81). This study demonstrated that a combination of alkaline and acidic EO water wash is equivalent to 200 mg/liter of chlorinated water wash for reducing populations of Salmonella Enteritidis and L. monocytogenes on shell eggs.

Science Direct-Application of electrolyzed oxidizing water to reduce Listeria monocytogenes on ready-to-eat meats. Friday, April 1st, 2005 K.A. Fabrizio, and C.N. Cutter, Department of Food Science, 111 Borland Laboratory, The Pennsylvania State University, University Park, PA 16802, USA

Experiments were conducted to determine the effectiveness of acidic (EOA) or basic electrolyzed oxidizing (EOB) water, alone or in combination, on ready-to-eat (RTE) meats to reduce Listeria monocytogenes (LM). Frankfurters or ham surfaces were experimentally inoculated with LM and subjected to dipping or spraying treatments (25 or 4 °C for up to 30 min) with EOA, EOB, and other food grade compounds. LM was reduced the greatest when frankfurters were treated with EOA and dipped at 25 °C for 15 min. A combination spray application of EOB/EOA also resulted in a slight reduction of LM on frankfurters and ham. However, reductions greater than 1 log CFU/g were not observed for the duration of the study. Even with a prolonged contact time, treatments with EOA or EOB were not enough to meet regulatory requirements for control of LM on RTE meats. As such, additional studies to identify food grade antimicrobials to control the pathogen on RTE meats are warranted

---

CONTROL OF POWDERY MILDEW BY SPRAYING THE ELECTROLYZED WATER IN HYDROPONICALLY GROWN STRAWBERRY ISHS Acta Horticulturae 559: V International Symposium on Protected Cultivation in Mild Winter Climates: Current Trends for Suistainable Technologies

Abstract:

Effect of spraying electrolyzed acid water and alkaline water obtained by electrolysis of KCl solution on the incidence of powdery mildew (Sphaerotheca macularis) is investigated in strawberry (Fragaria x ananassa Duch. cv. Nyoho) grown in hydroponics. Free effective chlorine concentration of the acid water at 0.5m from spraying nozzle ranged between 5.7 and 15.6 mg/l. pH and oxidation-reduction potential of the acid water ranged from 2.2 to 2.3 and from 1100 to 1140 mV, and those of the alkaline water ranged from 12.3 to 12.6 and from 30 to 280 mV, respectively. The treatment plots were 1. No spraying (NSp), 2. Acid water spraying (Ac), 3. Alkaline water spraying 30 minutes after acid water spraying (Ac+Al), and 4. Agricultural chemicals spraying (Chem). Those electrolyzed waters were sprayed onto the leaves and petioles once every week, and chemicals were sprayed once every two or three weeks.

The results indicate that the acid water spraying and acid + alkaline water spraying can control the powdery mildew in strawberry and can reduce the use of chemical fungicide in protected cultivation.

Science Direct-Toxicity of mixed-oxidant electrolyzed oxidizing water to in vitro and leaf surface populations of vegetable bacterial pathogens and control of bacterial diseases in the greenhouse. Tuesday, December 28th, 2004. Ken Pernezny, , Richard N. Raid, Nikol Havranek and Jairo Sanchez. IFAS, Everglades Research and Education Center, University of Florida, 3200 E. Palm Beach Road, Belle Glade, FL 33430, USA

Efficacy of electrolyzed oxidizing water for the microbial safety and quality of eggs. Poultry Science 2004 Dec;83(12):2071-8. Bialka KL, Demirci A, Knabel SJ, Patterson PH, Puri VM. Department of Agricultural & Biological Engineering, The Pennsylvania State University, University Park, Pennsylvania 16802, USA.
During commercial processing, eggs are washed in an alkaline detergent and then rinsed with chlorine to reduce dirt, debris, and microorganism levels. The alkaline and acidic fractions of electrolyzed oxidizing (EO) water have the ability to fit into the 2-step commercial egg washing process easily if proven to be effective. Therefore, the efficacy of EO water to decontaminate Salmonella Enteritidis and Escherichia coli K12 on artificially inoculated shell eggs was investigated. For the in vitro study, eggs were soaked in alkaline EO water followed by soaking in acidic EO water at various temperatures and times. Treated eggs showed a reduction in population between > or = 0.6 to > or =2.6 log10 cfu/g of shell for S. Enteritidis and > or =0.9 and > or =2.6 log10 for E. coli K12. Log10 reductions of 1.7 and 2.0 for S. Enteritidis and E. coli K12, respectively, were observed for typical commercial detergent-sanitizer treatments, whereas log10 reductions of > or =2.1 and > or =2.3 for S. Enteritidis and E. coli K12, respectively, were achieved using the EO water treatment. For the pilot-scale study, both fractions of EO water were compared with the detergent-sanitizer treatment using E. coli K12. Log10 reductions of > or = 2.98 and > or = 2.91 were found using the EO water treatment and the detergent-sanitizer treatment, respectively. The effects of 2 treatments on egg quality were investigated. EO water and the detergent-sanitizer treatments did not significantly affect albumen height or eggshell strength; however, there were significant affects on cuticle presence. These results indicate that EO water has the potential to be used as a sanitizing agent for the egg washing process.

Susceptibility of Penicillium expansum spores to sodium hypochlorite, electrolyzed oxidizing water, and chlorine dioxide solutions modified with nonionic surfactants. Journal of Food Protection. 2006 Aug;69(8):1944-8. Okull DO, Demirci A, Rosenberger D, LaBorde LF.

The disinfection of the surgery department of a garrison hospital using a neutral anolyte. Voen Med Zh. 1999 Sep;320(9):56-8. Chueva IM. Mistriukov VV, Mikhailov SI. [Article in Russian]

Methods and resources of sterilization and disinfection in medicine. Abstracts (28 September - 2 October 1992). Russian Conference.

Electrochemistry. Electrochemistry and Medicine. // Itogi nauki I tekhniki. VINITI. Ed. Polukarov YM, vol. 31, 1990. Article in Russian.

Bibliography

Ion-exchange Membrane Separation Processes, by Strathmann, H., *Membrane Science and Technology Series, 9*, Elsevier, Amsterdam, 2004. 360 pages. This book is the bible of technical processes of ionization. It is written at the PhD level and is meant for the scientific minded only. An in-depth understanding of thermodynamic and physicochemical processes is required. This book discusses the most relevant aspects of ion-exchange membranes, such as those used in water ionization. It discusses historical development, as well as the structure of ion-exchange membranes and their various processes.

Fantastic Voyage by Ray Kurzweil and Terry Grossman, MD.

The Secret Life of Water, by Masaru Emoto. 2005. Publisher: Atria. 224 pages

The Hidden Messages in Water, by Masaru Emoto. 2004; Publisher: Beyond Words Publishing. 160 p.

Your Body's many Cries for Water, by F. Batmanghelidj, M.D. Publisher: Global Health Solutions, Inc. 3rd Edition. 2001. 196 p.

Secrets of an Alkaline Body. The New Science of Colloidal Biology, by Annie Padden Jubb and David Jubb. Publisher: North Atlantic Books. 2004.

Consumer's Guide to Purchasing a Water Ionizer, by Bob McCauley, CNC, MH. Publisher: SE, Inc. 2009. 31 p.

Achieving Great Health - How Ionized Water, Spirulina, Chlorella, Raw Foods Can Make You Healthier than Ever Imagined in 90 Days or Less, by Bob McCauley, CNC, MH. Publisher: SE, Inc. 2005. 211 p.

Honoring the Temple of God – A Christian Health Perspective, by Bob McCauley, CNC, MH. Publisher: SE, Inc. 2008. 105 p.

Confessions of a Body Builder, by Bob McCauley, CNC, MH. Publisher: SE, Inc. 2000. 104 p.

Other Books that Contain References about Ionized Water:

Handbook of Drinking Water Quality, by John De Zuane. Publisher: John Wiley and Sons. 1997. Pages: 592

Pain Free With Far Infrared Mineral Therapy: The Miracle Lamp, by Kara Lee Schoonover. Publisher: iUniverse ISBN: 0595272630. 2003.

Sensual for Life: The Natural Way to Maintain Sexual Vitality, by George L. Redmon. Publisher: Kensington Books ISBN: 0758201389. 224 pgs. 2003.

Natural Choices for Fibromyalgia: Discover Your Personal Method for Pain Relief, by N. D. Oelke, Jane Oelke. Publisher: Natural Choices, Inc. ISBN: 0971551200. 2001. 146 pgs.

Light Through Food: A Reference Guide to Thriving in the Energy of the Fourth Dimension, by Donna Boynton (Channeler), Donna Boynton, Angela McLean. Publisher: Trafford Publishing ISBN: 1412068495. 2005. 100 pgs.

Qi Energy for Health and Healing: A Comprehensive Guide to Accessing Your Healing Energy, by Mallory Fromm. Publisher: Avery. ISBN: 1583331573. 2003. 219 pgs.

"Ionized Water is also very helpful for alleviating the side effects of anesthesia."

Romo: My Life on the Edge, by Bill Romanowski, Published by William Morrow. ISBN: 006075866X. Publisher: HarperTorch. 2005.

A Complete Guide To Chi-Gung, By Daniel Reid. ISBN: 1570625433 Publisher: Shambhala. 2000. 326 pgs.

Secrets of an Alkaline Body. The New Science of Colloidal Biology, by Annie Padden Jubb and David Jubb. Publisher: North Atlantic Books. ISBN: 1556434812. 275 pgs. 2004.

Timeless Secrets of Health and Rejuvenation, By Andreas Moritz. ISBN: 097657151X; pg. 496. 2005. Publisher: Ener-Chi Wellness Center, LLC

Reverse Aging, by Sang Whang. ISBN: 0966236327, Published by JSP Publishing. 2001.

Fantastic Voyage: Live Long Enough to Live, by Ray Kurzweil, Terry Grossman **ISBN:** 0452286670. 464 pages.

Appendix 6
Testimonials

I have received hundreds of testimonials over the years, both verbal and written. Provided anonymously, here is but a small sampling of some of the best ones.

I just started taking your chlorella and spirulina and also the water - haven't had the ionized in a few months. I'm very impressed with the results in just a few days. My joint pains are about 80% better and I feel much calmer and more alert. I've been drinking alkaline water and taking the spirulina and chlorella and eating lots of raw food. I'm getting great results! I've noticed, though, that eating just fruit in the morning and for snacks seems to make me very nervous. I have a bad case of Candida and a horrible yeast infection for 9 years now, can't get rid of it. I take colloidal silver but it doesn't help much. I also don't lose weight which I expected eating all the raw food. A doctor recently did a blood analysis and said I had low thyroid. This morning I had just vegetable juice with the green food and I noticed a world of difference - I felt much calmer, more alert, peaceful and nourished. I totally believe in this program and I know it can work.
Lucille F

Letter written to a debunker by one of my customers:
Whilst you are of course entitled to your opinion, maybe you should be little more open minded than what you appear to be. While I have never had any major health concerns I had been plagued by Tinnitus for a number of years. I have visited several specialists in an attempt to gain some relief. What a waste of time! The "so called" experts have no idea what causes it, let alone has a cure. One day I read that Tinnitus may have its origins as a result of an over acidic physiology and that alkalizing the body may help to reduce its severity. Skeptical I was to say the least, but I thought what have I got to lose so I began to explore ways of how I could reduce my body's acidity.

That is how I stumbled upon water ionizers and so I purchased the top of the line. Over several months I changed nothing else about my habits other than to drink ionized water, but before I began I tested my urine using a pH test kit several times over 2 days and returned a average reading of 5.5-6.0. After the first 2 days of drinking the water I began to notice that when I woke up in the morning my joints were very sore, so sore in fact that on the 4th day I could hardly get up out of bed. Later in the day I developed a headache which lasted for nearly 24 hrs. I have NEVER had a severe headache in my life, much less one as severe as this. Gradually over the next 2-3 days all the symptoms went away. It was after about 2 weeks that one morning I noticed something. I woke up feeling like I had the best night sleep I have ever had. I felt great, the whole day I felt like I had energy to burn, but that's not all, my Tinnitus which was always with me was hardly noticeable. I couldn't believe it! It's now been just over 2 months since I commenced drinking ionized water. My urine samples are now registering 7.5-8.0. So you tell me, is it all in my imagination? Because if it is you had better tell my 11 YO German Shepherd that has been plagued with chronic ear infection and Candida overgrowth for years. Despite repeated trips to several vets and countless antibiotics and antifungal medication, NOTHING has worked. I started Roxy on ionized water at the same time as myself. Her ear has cleared up her skin is back to normal and she is like a pup once again. Like I said, maybe it's just in her mind huh? Tell me what would happen if I drank 2 liters of Coke per day? I know because I had a friend that did just that for years. He now has severe stomach ulcers, his pancreas is shot to pieces and he has damaged his liver and suffers from Diabetes. If Coke with a low pH can do this sort of damage then is it not feasible that drinking water with a high pH could have the reverse effect and be beneficial? Do you mean to tell me that all the literature and research being done in Japan is bunkum? Remember Columbus and what they thought of him!

Allan Q
July 19, 2005

I ordered the water ionizer and it has really helped me drink more (life preserving) water. Thank you.

Kathleen G
September 26, 2005

I bought a water ionizer about two years ago. I like it very much along with my whole family, if I lost it I would want to replace it as quick as I could.

Randy T

My health challenge: stomach problems and tiredness
How has your health changed: my awareness of needing to drink more water has drastically risen this last year simply because we have the ionizer. Also, the water tastes better and therefore I drink close to my daily need. I feel much more hydrated. My eyes aren't puffy anymore and my skin is tighter. In the mornings when I wake up, I don't have an upset stomach and feel more rested! Viva la Ionizer. Thanks.
Ignatius T
January 5, 2006

Dear Bob, Patti and Ryan,
I wanted to express my thanks to all of you for providing such a fantastic opportunity to better my health. I have suffered from severe acidic reflux and gastritis since I was about 12 years old. I can honestly say I have taken nearly every prescription available for acid reflux disease and none of them have ever been effective. After visiting a second gastroenterologist and still not getting the results I anticipated I assumed that I would live with stomach pain my entire life. One day while picking up ionized water, I mentioned to Patti that my stomach was really hurting that day. She said here, and handed me a single probiotic pill. I was obviously skeptical, but I took it, and the next day realized my stomach didn't hurt as bad. So, I bought a bottle of probiotics and started taking it everyday. After a week I had to call Patti and tell her, I had gone an entire week without a single stomachache, this was unheard of for me! I have also been taking the spirulina and chlorella as well as drinking the ionized water, and I can't believe how much more energy I have, but best of all, no stomachaches. I haven't even had a cold this winter! I was also recently exposed to the Norwalk virus, via the contamination at an Italian restaurant in Lansing. I did not get sick, and I firmly believe it was due to the [Bob McCauley's] regimen that has built up my immune system. Bob, I would also like to thank you for implementing the rent-to-own plan for the water ionizers. This is a great opportunity to make great health available to everyone, regardless of his or her financial situation. Words cannot express my gratitude for everything you have done for my personal health. Please keep up the great work!

Sincerely yours,
Meghan M
February 3, 2006

Health challenge: winter skin fix, health issues
How has your health changed: my skin has become very soft, not itchy. I have not been sick since starting the Ionized Water and Celtic sea salt. I would encourage everyone to try this. This has really changed my life.
Ed F
February 17, 2006

The book, "Achieving Great Health" was wonderful! I have gone meatless, soft drink free and use only structured, ionic, alkalized water with my Far-Infra-Red Sauna. I do feel much better but I still have a whole lot of body pain.
Randolph Y
February 21, 2006

My health challenge: dry skin on face. How has your health changed: using acid water on my face has taken the dry scaly, skin off in only a weeks time.
Amy Y
February 25, 2006

My health challenge: it is very dry here in Las Vegas
How your health has changed: Ionized Water is delicious! I have astounded many friends with the taste and health benefits of better water. They all report how much more moisture is in their bodies and skin.
John M
February 28, 2006

Hope all is well with you and yours! My name is Michelle and I'm writing you from Buffalo, New York. I wanted to let you know that I purchased a ionizer from you almost 2 yrs. ago and have been impressed with my ionic water! It has become a lifestyle for me.
Michelle H
April 4, 2006

Glossary of Terms

Acid (Acidic): A substance that yields hydrogen ions when dissolved in water. When the body accumulates hydrogen ions, it becomes acid. Any pH below 7.0 is acid. The lower the number, the stronger the acid concentration is.

Acidify: To make any substance more acid by adding hydrogen ions.

Anode: A positively charged electrode, as of an electrolytic cell, storage battery, or electron tube. The electrode at which oxidation occurs.

Anabolic: Processes that are constructive in their nature. The phase of metabolism in which simple substances are synthesized into the complex materials of living tissue.

Electrolysis: Changing the chemical structure of any compound through the use of electrical energy.

Alkaline: A substance with a pH higher than 7.0. Any substance having properties where there are more hydroxyl ions than hydrogen ions. Anything that is alkaline is considered a base.

Alkalize: To make more alkaline by adding hydroxyl ions to the liquid.

Antioxidant: Any chemical compound, liquid or substance that inhibits oxidation. Any substance that contains large amounts of electrons such as raw foods and *Ionized* Water would be considered an antioxidant. The centerpiece of *Ionized Water* is its antioxidant properties.

Catabolic: Processes that are destructive in their nature. Catabolism is the metabolic process that breaks down molecules into smaller units.

Cathode: The negatively charged electrode, from which electrons are emitted and to which positive ions are attracted.

Chlorella Pyrenoidosa: A green single-celled micro-algae. The most powerful whole food in the world, which possesses powerful antioxidant, immune-building and anti-cancer properties. Also known for removing heavy metals from the body. There are other strains of chlorella, such as *vulgaris*, that are not as powerful or concentrated in nutrients.

Deionized Water: Purified Water that is the exact opposite of *Ionized Water*. Water that has been treated to remove minerals. This water should not be consumed.

Detoxification (detoxify): To counteract or destroy any substance that is toxic to the body. To removes toxins, or poisons, from the body. *Ionized Water* is extremely detoxifying because it effectively removes toxins from the body.

Electrode: The electrodes that are used to ionize water are comprised of 100% Titanium and coated with 100% Platinum. Other elements have been experimented with over the years, however, no better combination of metals has been found to produce strong alkaline and acid Ionized Water. Virtually all water ionizer manufacturers use exclusively 100% Titanium and Platinum. The terms "medical grade" and "surgical grade" Titanium and Platinum have been used to market water ionizers, however it is only necessary to know that they are indeed 100% pure.

Electron: The lightest electrically charged subatomic particle in existence. It is one of several small elementary particles that circle the nucleus of an atom. An electron has the same mass and amount of charge as a positron, which is positively charged. Electrons are negatively charged. The average person is starved for electrons.

Fluorine: Fluorine is the most reactive element in existence, attacking such inert materials as glass. Fluorine reacts with considerable violence with most hydrogen-containing compounds.

Free Radical: An atom or group of atoms that has at least one unpaired electron and is therefore unstable and highly reactive. In animal tissues, free radicals can damage cells and are believed to accelerate the progression of cancer, cardiovascular and other diseases.

Halogen: An element of a closely related group of elements consisting of fluorine, chlorine, bromine, and iodine. They combine readily with most metals and nonmetals to form a variety of compounds and never occur uncombined in nature. All are highly reactive oxidizing agents with valence 1 (for fluorine, the only valence).

Hydrogen Ion: A positively charged species of chemical, symbol H+. The ionized form of the hydrogen atom. Also known as a free radical. The buildup of hydrogen ions in the body leads to accelerated aging and creates an environment for disease to flourish.

Hydroxyl Ion: The anion having one oxygen and one hydrogen atom, denoted as OH^-. An oxygen molecule that carries with it an extra electron that can be donated to a free radical.

Ion: An atom or a group of atoms that has acquired a net electric charge by gaining or losing one or more electrons.

Ionize: A process that results in the gain or loss of electrons from an atom. To convert totally or partially into ions.

Ionization: The physical process of converting an atom or molecule into an ion by changing the difference between the number of protons and electrons. The **self-ionization of water** is the chemical reaction in which two water molecules react to produce a hydronium (*Acid Ionized Water*: H_3O^+) and a hydroxyl ion (*Alkaline Ionized Water*: OH^-).

Ionized Water: Water produced through the process of electrolysis and separated by a membrane into *Acid* and *Alkaline Ionized Water*. The alkaline water is an antioxidant, the acid water is for external use only.

Mineral Water: Water that contains large amounts of dissolved minerals such as calcium, sodium, magnesium, and iron.

Naturalist: One who pursues health using the most natural approach, including *Ionized Water*, Spirulina and Chlorella, Probiotics, Raw Foods, Exercise and cultivating a positive mental Attitude or Spiritual life. I consider myself someone who studies and promotes health in the most natural way conceivable.

ORP (Oxidation Reduction Potential): The ability or potential of any substance to reduce the oxidation of another substance which would be considered an antioxidant. The tendency of chemical substances to acquire electrons from other substances. Also known as redox potential.

Oxidize: To combine any substance with oxygen, which results in the loss of electrons from that substance. Oxidation of the human body results in accelerated aging. Hydrogen ions oxidize substances, such as those found in *Acid Ionized Water*.

pH (potential of hydrogen): The measurement of the acidity or alkalinity of a solution. A measurement of the electrical resistance between positive and negative ions. A pH of 7 is neutral. Any pH above 7 is alkaline. Any pH below 7 is acid. pH increases or decreases as the concentration of hydrogen ions increases or decreases. The more hydrogen ions, the more acid the solution becomes and the pH decreases. The pH scale ranges from 0 to 14, where 0 is absolute acid.

Purified Water: The result of removing all minerals from water through the mechanical processes of reverse osmosis, de-ionization or distillation. It is a pure chemical substance, H_2O. This water should not be consumed.

Redox: Oxidation-reduction. (See ORP)

Spirulina Platensis: A blue-green algae that is one of the two most powerful whole foods in the world. It is an extremely high-energy food. A microscopic freshwater plant, an aquatic micro-vegetable/organism composed of transparent bubble-thin cells stacked end-to-end forming a helical spiral filament. Spirulina has been consumed for thousands of years. It contains Gamma Linolenic Acid (GLA), well-known for relieving arthritic and rheumatoid conditions.

Spring Water: Water bottled from a source which flows out of the ground. Spring water has a great potential to be contaminated by pollution because it is drawn close to the surface.

Toxin: Any substance that is poisonous to the body. Toxins are things that serve no useful purpose in the human body and therefore should be removed. *Ionized Water* is an excellent way to help remove toxins for the body. All disease lives on toxins.

Valence: The combining capacity of an atom or radical determined by the number of electrons that it will lose, add, or share when it reacts with other atoms. The number of binding sites of a molecule, such as an antibody or antigen.

Zeta Potential: The ability for objects, such as blood cells to repel one another so they do not coagulate. The most important factor that affects zeta potential is pH. A zeta potential can only be determined in the context of its environment, using factors such as pH, ionic strength or concentration of any additives. A toxin can be defined as a substance that possesses the ability to overcome the zeta potential of the colloidal solution that it is in, meaning that it can not be removed from that solution except by exterior forces, i.e., detoxifying substances.

Index

About the Author

Bob McCauley, CNC, MH (Robert F., Jr.) was raised in Lansing, Michigan and attended Michigan State University (BA, 1980 in Journalism). He has traveled extensively, both domestically and abroad, visiting over 32 countries. He published Confessions of a Body Builder: Rejuvenating the Body with Spirulina, Chlorella, Raw Foods and *Ionized Water* (2000), Achieving Great Health (2005), Twelve (2007), Honoring the Temple of God (2008) and Consumer's Guide to Purchasing a Water Ionizer (2009). He considers himself a *Naturalist*, meaning that he pursues health in the most natural way possible. He studies and promotes nature as the only way to true health. He originated the Raw Food Pyramid and is considered a leader in the promotion of Spirulina And Chlorella. From 2002-2004 he hosted the radio program *Achieving Great Health,* which was heard by thousands of people each day.

With the help of his father, Dr. Robert F. McCauley, Sr.[93] (Doctorate in Environmental Engineering, MIT, 1953) they started Spartan Water Company in 1992, which sold vended water machines in supermarkets. Robert Jr. founded Spartan Enterprises, Inc. in 1993. He is a Certified Water Technician with the State of Michigan. He is also a Type II Public Water Supply Specialist. Bob often lectures and offers seminars on his Six Component Natural Health Protocol.

Bob is a Certified Nutrition Consultant and a Master Herbalist. He is also a 3rd Degree Black Belt and Certified Instructor of *Songahm Taekwondo* (American Taekwondo Association). He stays young by running, practicing Chi Gong and following his rules for Great Health laid out in this book. He also enjoys wall and rock climbing, not to mention breaking a brick now and then. This is his third book on health. Twelve is Bob's first collection of short fiction.

My Photos

1. Phil; 2. Rose and Phillip; 3. Me, Dan and Randy Couture, 3x Heavyweight UFC Champion. Randy loves Ionized Water; 4. Me and Bill Wang, Yi Shan; 5. Daniel & Me: Ronald Reagan Airport; 6. Daniel – Cross Country Regional's 2009; 7. Rose next to J. Bond; 8. Rose and Bob – 2009, Tai Chung, Taiwan.

More Books on Ionized Water

Written by

Bob McCauley, CNC, MH

ENDNOTES

---

1 **Ninth Amendment - Unremunerated Rights.** The enumeration in the Constitution, of certain rights, shall not be construed to deny or disparage others retained by the people.
Tenth Amendment - Reserved Powers. The powers not delegated to the United States by the Constitution, nor prohibited by it to the States, are reserved to the States respectively, or to the people.

2 Written by Bob McCauley. Published by Spartan Enterprises, Inc. Publication date: September 20, 2000. ISBN: 0-9703933-1-8. 114 p.

3 Written by Bob McCauley. Foreword by Peter Ragnar and Introduction by Gabriel Cousens, MD. Published by Spartan Enterprises, Inc. Publication date: April 10, 2005. ISBN: 0-9703933-6-9. 212 p.

4 If a person is too alkaline, which is extremely uncommon, it is because of their diet and the imbalances it has imposed on the body.

5 **ORP** stands for *Oxidation-Reduction Potential* or *Redox Potential).* It is a measurement of the ability of one substance to oxidize another. Potential refers to electrical potential, which is measured in millivolts. Oxidation means to combine with oxygen.

6 To Err Is Human: Building a Safer Health System. Kohn L, ed, Corrigan J, ed, Donaldson M, ed. Washington, DC: National Academy Press; 1999.

7 Ibid.

8 Increase in US medication-error deaths between 1983 and 1993. Phillips DP, Christenfeld N, Glynn LM. Lancet. 1998 Feb 28;351(9103):643-4.

9 In 1887, Swedish chemist Svante Arrhenius deduced that in dilute solutions electrolytes are completely dissociated into positive and negative ions. He applied this theory to acids and bases, arguing that acids dissociate to produce H^+ ions, and bases, OH^- ions. He received the 1903 Nobel Prize in Chemistry.

10 Chlorella is a green algae. The fiber in Chlorella is renowned for removing heavy metals and other synthetic toxins from the body. One of the absorbing substances in Chlorella fiber is sporopollenin, a naturally occurring carotene-like polymer that is extremely resistant to degradation. Chlorella's cell wall fiber binds with toxic

substances such as mercury and effectively removes them from the body through the bowels. Chlorella also readily removes lead, uranium, dioxins and cadmium from the body.

[11] An extract of brown kelp seaweed shown to remove heavy metals. It is also a powerful immune and cardiovascular builder.

[12] Toxins excreted due to FIR wave penetration include heavy metals, pesticides, *trans* fats, radiation, drugs and other petroleum-based chemicals, as well as fat itself. None of these toxic substances appear in typical sweat, only in sweat produced from a FIR sauna, in part because FIR waves penetrate 2" – 4" into the flesh to warm the body from within.

[13] For instance, never consume *Acid Ionized Water* because it is an oxidant that contains free radicals.

[14] "Theoretical Study of the 6 (H_2O) Cluster," C.J. Tsai and K.D. Jordan, *Chemical Physics Letters* 213, 181-88 (1993).

"Theoretical Study of Small Water Clusters: Low-Energy Fused Cubic Structures for (H_2O)n, n=8, 12, 16 and 20," C.J. Tsai and K.D. Jordan, *Journal of Physical Chemistry* 97, 5208-10 (1993).

This research is supported and validated by the National Science Foundation.

[15] A *dyne* is an extremely small measurement of force. 100,000 *dynes*= 0.224 lbs of force.

[16] "Water molecular arrangement at air/water interfaces probed by atomic force microscopy", by O. Teschke and E. F. de Souza. *Chem. Phys. Lett.* **403** (2005) 95-101. (b)

"Water molecule clusters measured at water/air interfaces using atomic force microscopy", O. Teschke and E. F. de Souza, *Phys. Chem.* **7** (2005) 3856-3865.

[17] The Sea Around Us, by Rachel L. Carson. 1951.

[18] A *Naturalist* is referred to in this book as one who pursues health in the most natural way possible. This includes *Ionized Water*,

Spirulina and Chlorella, Probiotics, Raw Foods, Exercise and cultivating a positive mental Attitude or Spiritual life. I consider myself someone who studies and promotes nature.

[19] *Alternatives Magazine*, Study of Great Britain's Olympic team by Dr. David G. Williams. August 2001.

[20] Jim Cantalupo McDonald's Corp. Chief Executive dies suddenly of heart attack at age 60 at a McDonald's convention in Orlando, Fla.

Charlie Bell, Cantalupo's successor died of colorectal cancer, age 44.

Coca-Cola Chairman Roberto C. Goizueta died of lung cancer , age 65.

Wendy's restaurant founder Dave Thomas died of liver cancer, age 69.

Jack Laughery, former CEO of Hardee's and former Chairman of its Board of Directors, died of pneumonia after a battle against lung cancer, age 71.

[21] Fantastic Voyage by Ray Kurzweil and Terry Grossman, MD.

[22] *Annals of the Rheumatic Diseases*, 2000; 59; 631-635.

[23] A pH of 2.7 is 10,000 times more acidic than the body's ideal pH of 7.0.

[24] "Consumption of soft drinks with phosphoric acid as a risk factor for the development of hypocalcaemia in postmenopausal women." Fernando, RM Martha and M. Djokic. 2003. *Journal Clinical Epidemiology.*

[25] Study done by the British Dental Association, reported in British *Dental Journal.* March, 2004.

[26] Routine tests revealed levels of up to eight parts per billion (ppb) in some soft drinks. Trace amounts found in Perrier Sparkling Natural Water™ in 1989 led to the withdrawal of more than 160 million bottles worldwide.

[27] "Consumption of Soft Drinks and Hyperactivity, Mental Distress, and Conduct Problems Among Adolescents in Oslo, Norway" Study conducted at the University of Oslo in Norway. Dr. Lars Lien, Nanna Lien, Sonja Heyerdahl, Magne Thoresen, and Espen Bjertness Am J Public Health, Oct 2006; 96: 1815 - 1820.

[28] Fantastic Voyage by Ray Kurzweil and Terry Grossman, MD.

[29] Dr. Valtin is the Andrew C. Vail Professor Emeritus, as well as former Chair of the Department of Physiology, at Dartmouth Medical School. His scientific and academic excellence has been recognized with several awards, including the Arthur C. Guyton Award for Distinguished Teaching in Physiology in 1994 and the Robert Berliner Award for Excellence in Renal Physiology in 1995.

[30] "Old 'rule' on water is all wet. Drink when thirsty, scientist concludes after much research", By Laura Beil. The Dallas Morning News. 08/19/2002

[31] Based on U.S. Department of Agriculture data soft drink consumption rose from 10.8 in 1946 to 49.2 gallons per capita, a 450% increase.

[32] Governor Arnold Schwarzenegger signed legislation making California the first state to ban soft drinks in schools. By July, 2007, students will be allowed only to buy water, milk, and some fruit and sport drinks that have limited sweeteners.

[33] Study done at Tata Memorial Hospital by Dr. Mohandas Mallath, et al. 2004.

[34] "Does aspartame cause human brain cancer?" Roberts HJ. *J Advanc M* 1991; 4 (Winter):231-241.

[35] "Carpal tunnel syndrome due to aspartame disease", Roberts HJ. *Townsend Letter for Doctors & Patients* 2000; 198 Nov: 82-84.

[36] US soldiers died of drinking too much water in too short a period in September 1999, January 2000, March 2001. *Military Medicine*, May

2002; 167: 432-434. The Army now recommends 1 – 1.5 quarts per hour and 12 quarts (3 gallons) per day.

[37] Ionic minerals are readily transported across the cell membranes of the human digestive tract. Because ionic minerals are charged, the body has to use less energy to absorb these minerals. These are also referred to as organic (ionic) and inorganic (non-ionic).

[38] Report by the *Earth Policy Institute*. February 2, 2006

[39] The sum of all inorganic and organic particulate material in water. There is a relationship between TDS and conductivity.

[40] Author of The Secret Life of Water and The Hidden Messages in Water. He is head of the I.H.M.General Research Institute Inc., the President of I.H.M.Inc.and the chief representative of I.H.M.'s HADO Fellowship.

[41] One of the world's foremost leading nutritionists, Norman Walker, PhD, wrote a number of books on juicing, nutrition and health that are still being printed and sought after. Around the age of 50, Dr. Walker developed severe cirrhosis (hardening) of the liver and nearly died. That led him to a diet of mainly salads, fruits, and vegetable and fruit juices. He died at age 118.

[42] Paul Bragg, Naturalist and health pioneer. Died at age 100 in an accident.

[43] Your Body's many Cries for Water, F. Batmanghelidj, M.D.

[44] Ibid.

[45] "Early Death Comes From Drinking Distilled Water," Zoltan P. Rona MD MSc.

[46] Paavo Airola, Ph.D., N.D. Internationally recognized nutritionist, naturopathic physician, award-winning author, and renowned lecturer. Regarded by many as a world-leading authority on holistic medicine and nutrition.

[47] "Early Death Comes From Drinking Distilled Water", Zoltan P. Rona MD MSc.

[48] Nannobacteria are approximately one-tenth the diameter of ordinary of bacteria. They are 30-100 nanometers (nm - billionths of a meter). They excrete calcium and are able to hide inside their own shell, safe from bodily defenses that recognize them as simple calcium in the bloodstream.

[49] "Early Death Comes From Drinking Distilled Water," Zoltan P. Rona MD MSc.

[50] Minerals contained in raw fruits and vegetables are ionic, meaning they have a negative charge, which makes them absorbable by the body. Once a food is cooked or processed, the charge is lost. Ionization provides minerals with a charge so they can be more easily absorbed by the body.

[51] "The Real Cost Of Bottled Water", Dr. Joseph Mercola, MD. 2006. www.mercola.com

[52] "Effects of alkaline Ionized Water on formation & maintenance of osseous tissues", by Rei Takahashi Zhenhua Zhang Yoshinori Itokawa. Study at the Kyoto University Graduate School of Medicine, Dept. of Pathology and Tumor Biology, Fukui Prefectural University)

[53] Chloramine is added to municipal water treatment plants because it is more stable and will not dissipate from water, ensuring disinfection of the water to all its consumers, regardless of how far they are from the water treatment plant. Chlorine, on the other hand, dissipates much more rapidly as it moves through the system. Chloramine, monochloramine (NH_2Cl), is formed by adding chlorine and ammonia to the municipal water source. Both carbon and KDF effectively filter both chlorine and chloramines.

[54] *Epidemiology.* 1998;9(1):21-28, 29-35.

[55] *Epidemiology.* May 1999;10:233-237.

[56] Epidemiology September 1999;10:513-517.

[57] Alum is naturally occurring and non-toxic because of its large molecular size that prevents it from being absorbed through skin pores. It should not be confused with its synthetic forms, such as aluminum chlorhydrate or aluminum zicarnium, that are readily absorbed through the skin. These toxic compounds are commonly found in carbonated soft drink cans, beer cans, baking-soda, pickles, aluminum foil, liquid antacids and antiperspirants.

[58] Dr. Charles Gordon Heyd, former President of the American Medical Association.

[59] Dr. Robert Carton, former EPA Scientist.

[60] *Journal of the American Medical Association*, Sept 18, 1943.

[61] Dr. Ludgwig Grosse, Chief of Cancer Research, U.S. Veterans Administration.

[62] Folk Medicine, J.C. Jarvis, M.D. Henry Holt & Co., 1958, HB, p. 136

[63] Romo: My Life on the Edge, by Bill Romanowski, a former all-pro NFL linebacker and a member of four Super Bowl winners.

[64] Anecdotes. "Accept a miracle instead of wit — " (1683–1765). Edward Young, English poet, best remembered for *Night Thoughts*.

[65] Molecule weight is the mass of one molecule of that substance, relative to the unified atomic mass unit u (equal to 1/12 the mass of one atom of carbon-12).

[66] Fantastic Voyage, by Ray Kurzweil. He is one of the world's leading inventors, thinkers, and futurists. He is a recipient of the National Medal of Technology and an inductee in the National Inventors Hall of Fame. Fantastic Voyage co-authored by Terry Grossman, MD, founder and medical director of Frontier Medical Institute, a longevity clinic in Denver, CO.

[67] Protons (p): 938.3 MeV (One million electron-volts). Electrons (e^-): 0.511 MeV.

[68] Secrets of an Alkaline Body. The New Science of Colloidal Biology, by Annie Padden Jubb and David Jubb.

[69] Secrets of an Alkaline Body. The New Science of Colloidal Biology, by Annie Padden Jubb and David Jubb

[70] "Ionized Water Explained," Dr. Hidemitsu Hayashi, Director, The Water Institute, Tokyo, Japan.

[71] Ibid.

[72] Ibid.

[73] Secrets of an Alkaline Body. The New Science of Colloidal Biology, by Annie Padden Jubb and David Jubb.

[74] "Ionized Water Explained," Dr. Hidemitsu Hayashi, Director, The Water Institute, Tokyo, Japan.

[75] Ibid.

[76] In the hydrogenation process, vegetable oil is placed under high pressure with hydrogen gas at 250 – 400° F for several hours in the presence of catalysts such as nickel, platinum, aluminum and other heavy metals that have been implicated in various brain diseases including bipolar and other dementia-related diseases. This process does not control where the hydrogen atoms are added to the unsaturated double bonds of the fat molecules. Randomly adding hydrogen atoms to polyunsaturated fats converts natural food components into numerous compounds, some that have never been seen before until partially hydrogenated fats were invented. This reveals how processing foods changes them into substances that no longer remotely resemble the natural whole foods that are recognized by the body. *Trans* fats have been scientifically linked to many health problems, including heart disease, diabetes, obesity, atherosclerosis, immune system compromise, reproductive and lactation issues and cancer.

---

[77] The vaccine protects against infection from four strains of human papillomavirus (HPV).

[78] *International Journal of Epidemiology.* October, 2001.

"Prevalence of overweight and obesity among U.S. children, adolescents, and adults, 1999-2002." A.A. Hedley, et al. *Journal of the American Medical Association*, 291(23): 2847-2850. 2004.

"Extent of overweight among U.S. children and adolescents from 1971 to 2000." D. Jolliffe. *International Journal of Obesity*, 28(4): 4-9. 2004.

[79] Although it has not been conclusively proven, it stands to reason that a highly oxygenated environment would be detrimental to cancer cells, which do not thrive in such an environment.

[80] "Ionized Water Explained," Dr. Hidemitsu Hayashi, Director, The Water Institute, Tokyo, Japan.

[81] Ibid.

[82] Alkalize or Die, by Dr. T. Baroody. Published by Holographic Health Inc; 9th ed edition (December 1991). 242 pages.

[83] Ibid.

[84] Greek tragic dramatist (484 BC - 406 BC)

[85] Nuclear magnetic resonance (NMR) is a physical phenomenon based upon the magnetic property of an atom's nucleus.

[86] **Zeta potential** is a measurement of the magnitude of the repulsion or attraction between particles. If **zeta potential** is low, **toxins** cannot be suspended in bodily fluids for elimination by the body and nutrients cannot be suspended for transportation to the cells so they can absorb and utilize them. Blood with low **zeta potential** is full of toxins. A toxin can be defined as a substance that possesses the ability to overcome the zeta potential of the colloidal solution that it is in.

[87] Secrets of an Alkaline Body. The New Science of Colloidal Biology, by Annie Padden Jubb and David Jubb.

[88] 60 to 70 million people affected by all digestive diseases (1996). Adams PF, Hendershot GE, Marano MA. Current estimates from the National Health Interview Survey, 1996. *National Center for Health Statistics. Vital Health Stat.* 1999; 10(200).

[89] "Bone mineral density in immigrants from southern China to Denmark. A cross-sectional study." Wang Q, et al. *Eur. J. Endocrinol.* 1996 / 134 (2) / 163-167.

[90] "Effect of calcium supplementation on bone mineral accretion in Gambian children accustomed to a low-calcium diet." Dibba B, et al. *Am J Clin Nutr* 2000. 71 (2). 544-9. , Aspray TJ, et al.

"Body measurements, bone mass and fractures: does the East differ from the West?" HO SC. *Clin. Orthopaed.* Related Res. 323 (1996): 75 – 80.

"Low bone mineral content is common but osteoporotic fractures are rare in elderly rural Gambian women*", J Bone Miner Res* 1996 / 11(7) / 1019-25.

"Osteoportic fracture rate, bone mineral density, and bone metabolism in Taiwan". J. Formosan *Med. Assoc*. 96 (1997): 75 – 80.

[91] Paracelsus. Swiss born physician. (1493-1541).

[92] The cavity in which the large intestine begins and into which the ileum opens. The large blind pouch forming the beginning of the large intestine. Also called the *blind gut.*

[93] Dr. Robert F. McCauley, Sr., was born and raised in the southwestern United States, Texas and New Mexico. Dr. McCauley served in W.W.II as a Lieutenant Colonel in North Africa and Central Asia. He received his Master's Degree from

Michigan State University in 1951 (Civil/Sanitary Engineering) and his Ph.D. from the Massachusetts Institute of Technology (MIT) in 1953 for his thesis on removing radioactive strontium from water. He earned his doctorate in Environmental Engineering in less than 2 years, one of the shortest doctoral studies in the history from MIT. He taught civil, sanitary and environmental engineering at Michigan State University for 18 years before retiring to run Wolverine Engineers & Surveyors of Mason, Michigan, for 17 years. With his business partner, George Young, they transformed the company into one of most respected and well known engineering firms in mid-Michigan. Dr. McCauley is credited with the invention and development of calcite coating for water main pipes, which keeps them from rusting. In 1992, he started Spartan Water Company, which provided vended water in supermarkets and was the precursor to Spartan Enterprises.

Dr. Robert F. McCauley, Sr.
1913-2000